Errata

Horst Sachs, *Characteristic polynomials in the theory of polyhedra*, in Graph Colouring and Applications (P. Hansen and O. Marcotte, eds.), CRM Proceedings & Lecture Notes, vol. 23, Amer. Math. Soc., Providence, RI, 1999, pp. 111–126.

Page 118, Line 13: Change "Lemma 10" to "Lemma 2".

Page 121: Figure 5 and Table 4 are correct as printed. Replace the text on this page by the following:

8. The Character Schemes of the Connected Subgraphs

Let G be a plane undirected graph, $\omega \in \Omega(G)$ and $\chi_\omega = \chi$. According to Proposition 2, the corresponding skew characteristic polynomial is

$$\widetilde{f}_\chi(G;x) = \widetilde{f}(G^\omega;x) = x^r(x^{2s} + a_2 x^{2s-2} + \cdots + a_{2s}).$$

By a classical theorem, the coefficient $a_{2\sigma}$ equals the sum of all principal minors of order 2σ of $\widetilde{A}(G^\omega)$. These minors are the determinants of the $\binom{v}{2\sigma}$ skew adjacency matrices of the induced directed subgraphs on 2σ vertices of G^ω. If H^ω is such a subgraph that is not connected, then the corresponding minor is the product of those minors that correspond to the components of H^ω. Therefore, it is desirable to know more about the character schemes of the connected subgraphs of G^ω.

THEOREM 2. *Let G be a plane undirected graph and $\widehat{G}$ a (not necessarily induced) connected subgraph of G. For any face $\widetilde{F}$ of $\widehat{G}$, let $\mathbb{F}^*(\widetilde{F})$ and $v(\widetilde{F})$ denote the set of those faces of G that are contained in $\widetilde{F}$, and the number of those vertices of G that lie in the interior of $\widetilde{F}$, respectively. Then, for any orientation ω of G*

Page 123: The "dodecahedron" line of Table 5, should appear as follows:

dodecahedron	$(x+2)^4 \times$ $x^4(x-1)^5 \times$ $(x-3)(x^2-5)^3$	$x(x-2)^5 \times$ $(x-3)^4(x-5)^4 \times$ $(x^2-6x+4)^3$	$(x^2+6)^4(x^2+1)^6$

Volume 23

CRM
PROCEEDINGS &
LECTURE NOTES

Centre de Recherches Mathématiques
Université de Montréal

Graph Colouring
and Applications

Pierre Hansen
Odile Marcotte
Editors

The Centre de Recherches Mathématiques (CRM) of the
Université de Montréal was created in 1968 to promote
research in pure and applied mathematics and related
disciplines. Among its activities are special theme years,
summer schools, workshops, postdoctoral programs, and
publishing. The CRM is supported by the Université de
Montréal, the Province of Québec (FCAR), and the
Natural Sciences and Engineering Research Council of
Canada. It is affiliated with the Institut des Sciences
Mathématiques (ISM) of Montréal, whose constituent
members are Concordia University, McGill University, the
Université de Montréal, the Université du Québec à
Montréal, and the Ecole Polytechnique. The CRM may be
reached on the Web at www.crm.umontreal.ca.

American Mathematical Society
Providence, Rhode Island USA

The production of this volume was supported in part by the Fonds pour la Formation de Chercheurs et l'Aide à la Recherche (Fonds FCAR) and the Natural Sciences and Engineering Research Council of Canada (NSERC).

1991 *Mathematics Subject Classification.* Primary 05C15, 05C90; Secondary 05C70, 68R10.

Library of Congress Cataloging-in-Publication Data
Graph colouring and applications / Pierre Hansen, Odile Marcotte, editors.
    p. cm. — (CRM proceedings & lecture notes, ISSN 1065-8580 : v. 23)
    Includes bibliographical references.
    ISBN 0-8218-1955-0 (alk. paper)
    1. Map-coloring problem Congresses. I. Hansen, P. (Pierre) II. Marcotte, Odile, 1951– .
III. Series.
QA612.18.G68   1999
514′.223–dc21   99-40201
                                                                                        CIP

Contents

Preface

From May 5 to May 9, 1997, the CRM hosted a workshop on Graph Colouring and Applications. The workshop was organized by Pierre Hansen and Odile Marcotte and was jointly sponsored by CRM and GERAD (Groupe d'études et de recherche en analyse des décisions). It brought together outstanding researchers in the fields of combinatorial optimization and graph theory, and included eleven lectures by invited speakers and seventeen contributed talks.

The opening lecture was given by Paul Seymour, from Princeton University, who presented the latest and most elegant proof of the famous four-colour theorem. This proof is the result of work carried out by Paul Seymour and his collaborators, several of whom were also attending the workshop. The lecture of Professor Carsten Thomassen, from the Technical University of Denmark, had a similar topic, i.e., algorithms for colouring graphs embedded in specific surfaces. Professors Claude Berge (from Paris) and Vasek Chvátal (from Rutgers University) gave lectures on perfect graphs, a topic closely related to the colouring problem. Professor Anthony Hilton (from Reading) lectured on total colourings of graphs, and Professor Adrian Bondy (from Lyon) on the relationship between colourings and orientations of graphs.

Graph colouring has many applications, some of whom arise from practical concerns and others from the natural or social sciences. Professor Horst Sachs (from the Technical University of Ilmenau, in Germany) presented results on invariant polynomials of polyhedra and their application to chemistry. Dominique de Werra (from the École Polytechnique Fédérale de Lausanne) gave a lecture on the application of graph colouring to the problem of register allocation, which arises during the compilation of computer programs. Fred Roberts (from Rutgers University) presented a variant of graph colouring used in the modelling of social roles, and related this concept to other important concepts in the mathematical models of the social sciences.

Finally, Mike Carter (from the University of Toronto) and Bjarne Toft (from Odense University, in Denmark) lectured on applications of graph colouring to timetabling. The last main lecture was given by Bjarne Toft, who educated and entertained his audience by discussing the life and correspondence of Julius Petersen, who gave his name to the famous Petersen graph! The themes of the contributed talks were similar to those of the main talks, and the workshop was attended by 46 researchers from eight countries (Canada, United States, France, England, Germany, Denmark, Switzerland and Israel).

The articles of this volume span a wide spectrum of topics related to graph colouring: enumeration of colourings (Walsh), list-colourings (Dror et al.), total colourings (Hamilton et al.), colourings and embeddings of graphs (Collins and Hutchinson), chromatic polynomials (Arrowsmith and Essam), characteristic polynomials (Sachs), chromatic scheduling (de Werra) and graph colouring problems related to frequency assignment (Walsh, Harary and Plantholt, Marcotte and Hansen). All the articles published in this volume were refereed. The proceedings also include a list of open problems suggested by the participants.

To conclude this introduction, we would like to express our heartfelt thanks to Dr. Luc Vinet, director of the CRM, for suggesting that the workshop be held and providing financial support. We would also like to thank the referees who reviewed the articles and Dr. Yvan St-Aubin, deputy director of the CRM, who took care of the refereeing process for our own article. Finally we thank Mr. Louis Pelletier for taking care of all aspects concerning the organization of the workshop.

Pierre Hansen
Odile Marcotte
Organizers of the workshop and
editors of the proceedings

Montréal, July 1999

Centre de Recherches Mathématiques
CRM Proceedings and Lecture Notes
Volume **23**, 1999

Chromatic Polynomials and mod λ Flows on Directed Graphs and Their Applications

D. K. Arrowsmith and J. W. Essam

ABSTRACT. We survey the problem of enumerating certain types of colourings of graphs which are dependent on orientation. Specifically, we consider colourings in which each of the $\nu = |V|$ vertices of a directed graph is coloured in one of the integer colours in $\mathcal{C} = \{0, 1, \ldots, \lambda - 1\}$ with the constraint that $(c_j - c_i)$ mod λ, when chosen as an element of $\mathcal{C}$, must be nonzero and even for all arcs (i, j), where c_i is the colour of vertex i. Similar enumerations can be made for the case of odd $(c_j - c_i)$ mod λ and they are distinguished by the parity of λ.

For even λ and a given undirected graph, the number of even colourings is independent of the directing and the problem is easily related to the standard colouring problem.

The main results concern the enumeration and connectedness properties of even colourings for odd λ when the colour difference on a particular arc has the fixed value β. There are corresponding enumerations and properties for mod λ flows. These restricted colourings and flows are applied in various ways. The key results describe the way in which the connectedness of directed graphs is derived from even colourings and flows. Thus they have relevance to the theory of directed percolation. The even flow enumerations have application in the theory of directed polymer networks. A generating function for the numbers of colourings with even and odd colour difference is seen as the partition function of a Potts model in which the colours are the states of spins attached to the vertices of the graph. For odd λ the model exhibits chirality.

Finally, we briefly consider the role of partition functions, and particularly those for the Potts Model, in producing knot invariants.

1. The Chromatic Polynomial

If the choice of attaching one of λ colours in $\mathcal{C} = \{0, 1, \ldots, \lambda - 1\}$ to each of the vertices of a graph $G = (V, E)$ is unrestricted, then the number of such colourings is counted by the polynomial $P(\lambda, G) = \lambda^\nu$, where $\nu = |V(G)|$ is the number of vertices of G. Furthermore, the number of ways in which the vertices of G can be coloured such that vertices adjacent to a common edge receive different colours, known as the number of *proper* colourings, is also wellknown to be counted by a

1991 *Mathematics Subject Classification.* Primary: 05C15; Secondary: 82B20.

We would like to thank Professor Christian Krattenthaler for drawing our attention to the connection between vicious walkers and plane partitions and also to thank the referees for suggesting several improvements to the paper.

This is the final version of the paper.

polynomial $P^+(\lambda, G)$ in λ, the *chromatic polynomial*. The graph G can contain loops but then $P^+(\lambda, G) = 0$. Moreover, the number of colourings is not changed if a multiple edge is reduced to a single edge. Every colouring of a graph G is equivalent to a proper colouring on a contracted graph G/G' obtained from G by contracting the edges E' which have identical colours on adjacent vertices and so we can write

$$(1.1) \qquad P(\lambda, G) = \sum_{E' \subseteq E} P^+(\lambda, G/G').$$

Möbius inversion gives the polynomial form of $P^+(\lambda, G)$ due to Whitney [**36**]:

$$(1.2) \qquad P^+(\lambda, G) = \sum_{E' \subseteq E} (-1)^{|E \setminus E'|} P(\lambda, G/G')$$

$$(1.3) \qquad\qquad = \sum_{E' \subseteq E} (-1)^{|E \setminus E'|} \lambda^{\nu(G/G')}.$$

In the following sections we shall impose further restrictions on the colourings and proper colourings. Equations (1.1) and (1.2) will also apply to these restricted colourings. Hence for each type of colouring considered in this paper, it is possible to restrict the enumeration to those either with or without the proper constraint. Also a proof of a polynomial property without the proper constraint will imply the same property for proper colourings and vice-versa.

1.1. Colour difference restrictions. We now consider $\mathcal{C}$ with mod λ addition as $\mathbf{Z}_\lambda$, the integers mod λ. Let $\mathcal{D}(G)$ be the set of directed graphs obtained by directing the edge set E of G in all possible ways. Let $c \colon V \to \mathcal{C}$ be a λ-colouring of G and let $H \in \mathcal{D}(G)$ have arc set $A(H)$. Associated with c and H is a mod λ colour difference $\delta c \colon A(H) \to \mathbf{Z}_\lambda$ where, if $a = (i, j) \in A(H)$, then $\delta c(a) = c_j - c_i$ mod $\lambda \in \mathcal{C}$. Clearly signed colour differences around a cycle sum to zero and this is also a possible defining property; that is an edge valuation is a colour difference (i.e. it supports a λ-colouring) iff its signed sum around every cycle is zero mod λ.

In the next section we pose a number of further colouring problems by imposing constraints on the image of the colour difference function δc. First of all we note that the more familiar proper and rooted colourings can be described in this way.

1.1.1. *Proper colourings.* Note that proper colourings are those for which $c_i \neq c_j$ for all $(i, j) \in A(H)$. Thus a λ-colouring c on $H \in \mathcal{D}(G)$ is *proper* if the colour difference δc satisfies $\delta c(A(H)) \subseteq \mathcal{C} \setminus \{0\}$.

1.1.2. *Rooted colourings.* A directed graph H becomes a rooted graph $H_{\bar{a}}$ by distinguishing an arc $\bar{a}$. A λ-colouring c on $H_{\bar{a}}$ is said to be *rooted* if the colour difference $\delta c(\bar{a})$ equals β, a fixed value in $\mathcal{C}$. The number of such rooted λ-colourings, $P_{\beta, \lambda}(H_{\bar{a}})$, is independent of the directing of H and is again counted by a polynomial in λ which for *nonzero* β is independent of β and given by

$$(1.4) \qquad P(\beta, \lambda, H_{\bar{a}}) = \lambda^{\nu(H)-1} \eta_{\bar{a}}(H),$$

where the *cocycle indicator* $\eta_{\bar{a}}(H)$ is 1 if H has a cocycle (or cut) containing $\bar{a}$ (i.e. $\bar{a}$ is not a loop) and is 0 otherwise.

1.2. The number of colour differences. For any given colour difference on $H \in \mathcal{D}(G)$, there are λ different associated colourings on any given connected component of G. Let the number of mod λ colour differences be denoted by $D_\lambda(H)$ then if $\omega(G)$ is the number of components in G

$$(1.5) \qquad\qquad D_\lambda(H) = \lambda^{-\omega(G)} P_\lambda(H).$$

This formula applies to all the colouring classes in this paper, in particular the number of unrestricted colour differences is counted by the polynomial

$$(1.6) \qquad\qquad D(\lambda, H) = \lambda^{\nu(H)-\omega(H)} = \lambda^{r(H)}$$

where $r(H)$ is known as the cocycle rank of H.

Consideration of the number of colour differences rather than the number of colourings brings out the duality relations with the number of flows to be considered later.

2. Colourings with Even and Odd Colour Differences

For either the proper or rooted cases described above we can impose further conditions on the λ-colourings. Given $H \in \mathcal{D}(G)$, consider those λ-colourings c for which the colour differences $\delta c(a)$ are even values of $\mathcal{C}$, $a \in A(H)$. The number of such "even" λ-colourings is denoted by $P_\lambda^{\mathrm{even}}(H)$. If the allowed values of the colour differences are taken to be odd or zero, then we have a further enumerator $P_\lambda^{\mathrm{odd}}(H)$. Crucially, for odd λ, the numbers of even and odd colourings are dependent on the directing of the underlying graph.

Moreover, the rooted analogues of these even and odd colourings have corresponding enumerators denoted by $P_{\beta,\lambda}^{\mathrm{even}}(H_{\bar{a}})$ and $P_{\beta,\lambda}^{odd}(H_{\bar{a}})$. The notation for colourings can be extended to colour differences of rooted graphs by using $D_{\beta,\lambda}^{\mathrm{even}}(H_{\bar{a}})$ and $D_{\beta,\lambda}^{\mathrm{odd}}(H_{\bar{a}})$. The properties of such functions are considered in detail in [5]. The origin of the problem addressed here lies in both the study of chiral Potts models and directed percolation [7] and decompositions of the chromatic polynomial [5].

The relationship (1.5) extends to rooted colourings and also even and odd colourings. Thus, for example, $D_{\beta,\lambda}^{\mathrm{even}}(H_{\bar{a}}) = \lambda^{-\omega(H_{\bar{a}})} P_{\beta,\lambda}^{\mathrm{even}}(H_{\bar{a}})$.

2.1. Odd λ. The enumerators of λ-colourings described in Section 1 were independent of the directing of the underlying graph G. It was shown that these colourings can be counted by polynomials of degree at most $r(G)+1$ and the cocycle indicator can be found by formally setting $\beta = 0$ and $\lambda = 1$ in (1.4) to obtain

$$(2.1) \qquad\qquad P(0, 1, G_{\bar{a}}) = \eta_{\bar{a}}(G).$$

We now describe analogous properties for even λ-colourings which for odd λ are directing dependent. In this case no explicit polynomial formulae have been found for a general graph and the proof of the polynomial property is much more difficult and requires an inductive argument.

We need the following definition. A set of arcs $b \subseteq A(H)$ is a *directed cut* of the graph H if the vertex set V has a nontrivial partition $[S, S']$ such that

(i) b is the set of arcs between S and S' and
(ii) the arcs of b are all directed from S to S'. Let $\chi_{\bar{a}}(H)$ be a directed cut indicator corresponding to $\eta_{\bar{a}}(G)$.

THEOREM 2.1. [5] *Let $H \in \mathcal{D}(G)$ have a rooted arc $\bar{a}$, then the number of colour differences, $D_{\beta,\lambda}^{\mathrm{even}}(H_{\bar{a}})$, for $\beta = 2m$ and $\lambda = 2n + 1$ with $m, n \in \mathbf{Z}^+$, may be obtained by evaluating a polynomial $D^{\mathrm{even}}(\beta, \lambda, H_{\bar{a}})$, in β and λ, having joint* *degree at most $r(H) - 1$, where $r(H) = \nu(H) - \omega(H)$ is the cocycle rank of H, with the property*

(2.2) $$D^{\mathrm{even}}(0, 1, H_{\bar{a}}) = \chi_{\bar{a}}(H),$$

where $\chi_{\bar{a}}(H)$ is the directed cut indicator for root arc $\bar{a}$. There exists a similar polynomial for β odd. Also, the values of $D^{\mathrm{even}}_{0,\lambda}(H_{\bar{a}}) = D^{\mathrm{even}}_{\lambda}(H^{\gamma}_{\bar{a}})$, where $H^{\gamma}_{\bar{a}}$ is the graph obtained by contracting the arc $\bar{a}$ of $H_{\bar{a}}$, are given by a polynomial $D^{\mathrm{even}}(\lambda, H^{\gamma}_{\bar{a}})$ of degree at most $r(H^{\gamma}_{\bar{a}})$ in λ such that $D^{\mathrm{even}}(1, H^{\gamma}_{\bar{a}}) = 1$. More generally, for $H \in \mathcal{D}(G)$, $D^{\mathrm{even}}(\lambda, H)$ is a polynomial in λ of degree at most $r(H)$ such that $D^{\mathrm{even}}(1, H) = 1$.

The proof of Theorem 2.1 is given in detail in [**5**]. The essential idea of the proof is to show by induction that $D^{\mathrm{even}}_{2m,2n+1}(H_{\bar{a}})$, where m, $n \in \mathbf{Z}^{+}$, can be enumerated by polynomials of degree at most $\nu(H) - \omega(H)$ in the variables m and n. An inductive step is set up which relates $D^{\mathrm{even}}_{2m,2n+1}(H_{\bar{a}})$ and $D^{\mathrm{even}}_{2m-2,2n+1}(H_{\bar{a}})$. Reversal of arcs on cuts are used to reduce the value of β from $2m$ to $2m - 2$. Thus the problem of evaluation of polynomials in the rooted case can be related to that of the unrooted case where the specific evaluations which give $D^{\mathrm{even}}(0, 1, H_{\bar{a}})$ can be made. The difference equation obtained together with a reversal-deletion-contraction rule (see Section 4.3.2) is sufficient to obtain the general polynomial properties given in the theorem.

LEMMA 2.2 (Directed cut reversal). [**5**] *Let b be a directed cut containing the root arc $\bar{a}$ and let H^{ρ}_{b} be the graph obtained by reversing all arcs of b other than $\bar{a}$. For odd λ, there exists a bijection between the even $\mod \lambda$ colour differences on $H_{\bar{a}}$ and H^{ρ}_{b} with difference β and $(\beta + 1)$ respectively in the root arc $\bar{a}$.*

The equivalence is obtained by increasing the colour difference on the arcs of b by one and then reversing the nonroot arcs of b. The resulting correspondence is given by $\delta c' = \delta c$ on the arc set $A(H) \backslash A(b)$ and for $a \in A(b)$, $a \neq \bar{a}$, $\delta c'(a^{\rho}) = \lambda - 1 - \delta c(a) \mod \lambda$ where $\delta c \in \mathcal{D}^{\mathrm{even}}_{\beta,\lambda}(H_{\bar{a}})$, $\delta c' \in \mathcal{D}^{\mathrm{even}}_{\beta+1,\lambda}(H^{\rho}_{b})$ and a^{ρ} is the arc a with reversed orientation. Thus

(2.3) $$D^{\mathrm{even}}_{\beta,\lambda}(H_{\bar{a}}) = D^{\mathrm{even}}_{\beta+1,\lambda}(H^{\rho}_{b}),$$

where $\beta + 1$ is evaluated $\mod \lambda$.

For odd λ, the various cases of the "even" and "odd" constraints and the parities of β are related by the following easy lemma, which shows that the counting polynomials for rooted colourings with odd colour difference can also be expressed in terms of those for colourings with even colour difference.

LEMMA 2.3. [**5**] *Let $H_{\bar{a}}$ be the rooted directed graph obtained from $H \in \mathcal{D}(G)$ with root arc $\bar{a}$. Let $H^{\rho}_{\bar{a}}$ be the rooted directed graph obtained from $H_{\bar{a}}$ by reversing the orientation of every arc of H except for the root arc $\bar{a}$. The graph H^{ρ} is obtained by reversing the orientation of every arc of H. For λ odd,*

(2.4) (i) $P^{\mathrm{odd}}_{\lambda}(H) = P^{\mathrm{even}}_{\lambda}(H^{\rho})$,

and

(2.5) (ii) $P^{\mathrm{odd}}_{\beta,\lambda}(H_{\bar{a}}) = P^{\mathrm{even}}_{\beta,\lambda}(H^{\rho}_{\bar{a}})$.

Moreover, if $(H_{\bar{a}})^{\rho}$ denotes $H_{\bar{a}}$ with all arcs reversed, then

(2.6) (iii) $P^{\mathrm{odd}}_{\beta,\lambda}(H_{\bar{a}}) = P^{\mathrm{even}}_{\lambda-\beta,\lambda}\big((H_{\bar{a}})^{\rho}\big)$, *where $\lambda - \beta$ is an element of $\mathcal{C}$ by reducing mod λ if necessary.*

2.2. Even λ. The problem of enumerating even and odd colourings is straight-forward when λ is restricted to be even [5]. Given H, $H' \in \mathcal{D}(G)$, a λ-colouring c on H with even colour difference δc is also even on H' since δc is the same on H and H' for all coherently oriented arcs and $\delta c(a) = \lambda - \delta c(a')$ mod λ, which preserves parity, when the arcs of the same edge $a \in A(H)$ and $a' \in A(H')$ are oppositely oriented. The same argument can be applied for λ colourings with odd colour differences which are also preserved by arc reversal. The number of colourings is therefore directing independent.

3.   mod λ Flows

The concept which is dual to mod λ colour differences is that of mod λ flows [35]. A mod λ flow on the directed graph $H = (V, A) \in \mathcal{D}(G)$ is a map $\phi \colon A \to \mathbf{Z}_\lambda$ such that

$$(3.1) \qquad \sum_{a \in A_v^+} \phi(a) = \sum_{a \in A_v^-} \phi(a),$$

in $\mathbf{Z}_\lambda$, for every vertex v, where the sets A_v^+, A_v^- are respectively the arcs oriented in and out of the vertex v. The number of such flows is given by a polynomial $F(\lambda, G)$ and depends only on the underlying graph G and not the directing H. Rooted flows on the directed graph $H_{\bar{a}}$ have the extra requirement that $\phi(\bar{a}) = \alpha$, a fixed value.

Furthermore, the structure dual to a directed cut is a *circuit* and is defined to be a directed cycle in which the arcs are coherently oriented. This includes the case of an oriented loop. The directed cut indicator can be related to the indicator for circuits $\pi_{\bar{a}}(H)$ which has value 1 if there is a circuit in H containing $\bar{a}$ and 0 otherwise. The existence of a directed cut containing $\bar{a}$ is equivalent to the nonexistence of a circuit containing $\bar{a}$ and so $\chi_{\bar{a}} = 1 - \pi_{\bar{a}}$.

We have the following theorem for flows which is dual to Theorem 2.1.

THEOREM 3.1. [5] *Let $H \in \mathcal{D}(G)$ have a rooted arc $\bar{a}$, then the number of flows $F_{\alpha, \lambda}^{\mathrm{even}}(H_{\bar{a}})$, for $\alpha = 2m$, $\lambda = 2n + 1$ with m, $n \in \mathbf{Z}^+$, may be obtained by evaluating a polynomial, $F^{\mathrm{even}}(\alpha, \lambda, H_{\bar{a}})$, in α and λ having joint degree at most $c(H) - 1$, where $c(H)$ is the cycle rank of H, with the property*

$$(3.2) \qquad\qquad F^{\mathrm{even}}(0, 1, H_{\bar{a}}) = \pi_{\bar{a}}(H).$$

There also exists a similar polynomial for α odd. Also, the values of $F_{0,\lambda}^{\mathrm{even}}(H_{\bar{a}}) = F_\lambda^{\mathrm{even}}(H_{\bar{a}}^\delta)$ are given by a polynomial $F^{\mathrm{even}}(\lambda, H_{\bar{a}}^\delta)$ of degree at most $c(H_{\bar{a}}^\delta)$ in λ such that $F^{\mathrm{even}}(1, H_{\bar{a}}^\delta) = 1$. More generally, for $H \in \mathcal{D}(G)$, $F^{\mathrm{even}}(\lambda, H)$ is a polynomial in λ of degree at most $c(H)$ such that $F^{\mathrm{even}}(1, H) = 1$.

REMARK 3.2. It should be noted that the particular evaluations in Theorems 2.1 and 3.1 of the interpolating functions D^{even} and F^{even} with $\beta = 0, \lambda = 1$ and $\alpha = 0, \lambda = 1$ respectively, which give the connectivity properties of a graph, are not the same as the combinatorial evaluations $D_{0,1}^{\mathrm{even}}(H)$ and $F_{0,1}^{\mathrm{even}}(H)$ which are both identically 1.

4. Applications

4.1. Connectedness of graphs and directed percolation.
Percolation theory was introduced by Broadbent and Hammersley [14]. Suppose that each arc of a directed graph H has probability p of being "open" and $1-p$ of being "closed". The central problem in directed percolation theory is to discuss the properties of the random set of vertices D which can be reached from a given vertex i by at least one directed path of open arcs. The main interest is in the case of infinite graphs in which case D has a positive probability, the percolation probability, of being infinite at or above a certain value of p, the critical probability. If we ignore the directings of the arcs of the path then we have the corresponding undirected percolation problem.

Another property which shows critical behaviour is the expected size $S_i(p, H)$ of D, known as the mean cluster size. $S_i(p, H)$ increases with p and becomes infinite at a critical value p_c of p depending on the graph. The value of p_c may be estimated by expanding $S_i(p, H)$ as a power series in p and using Padé approximant methods. To determine such an expansion we use the relation

$$(4.1) \qquad S_i(p, H) = \sum_{j \in V} C_{ij}(p, H),$$

where the *pair connectedness* $C_{ij}(p, H)$ is the probability of an open path from i to j. This may be expanded as a polynomial in p as follows.

By inclusion-exclusion we have the following expansion [2, 3],

$$(4.2) \qquad C_{ij}(p, H) = \sum_{S \subseteq \mathcal{S}_{ij}(H)} \mathrm{pr}(S)(-1)^{|S|+1}$$

where $\mathcal{S}_{ij}(H)$ is the collection of directed paths from i to j and $\mathrm{pr}(S)$ is the probability that at least the paths in the subset S are open. By noting that $\mathrm{pr}(S) = p^{|A_S|}$, where A_S is the union of the arc sets of the paths in S, and collecting together the terms of the above sum for which $A_S = A' \subset A$, we obtain a subgraph expansion which takes the form

$$(4.3) \qquad C_{ij}(p, H) = \sum_{A' \subseteq A} \vec{d}_{ij}(H') p^{|A'|},$$

where $H' = (V, A')$ and the corresponding coefficient $\vec{d}_{ij}(H')$ is solely a function of the subgraph H', known in the literature as the "d-weight", [7], of H'. By taking $p = 1$, and using Möbius inversion we have

$$(4.4) \qquad \vec{d}_{ij}(H) = \sum_{A' \subseteq A} (-1)^{|A \backslash A'|} \pi_{ij}(H'),$$

where $\pi_{ij}(H') = C_{ij}(1, H')$ is 1 if there a directed path in H' from i to j and zero otherwise.

For percolation on an undirected graph $G = (V, E)$ the "d-weight" in the expansion corresponding to (4.3) is denoted by $d_{ij}(G')$ and is given by

$$(4.5) \qquad d_{ij}(G) = \sum_{E' \subseteq E} (-1)^{|E \backslash E'|} \gamma_{ij}(G'),$$

where the undirected connectedness indicator $\gamma_{ij}(G')$ is 1 if there a path in G' from i to j and zero otherwise. Thus the connectedness of the subgraphs plays a major part in the calculation of the d-weights required to obtain the pair-connectedness.

Let $H_{\bar{a}}^+$ be the graph obtained from H by adding the "external" arc $\bar{a} = [j, i]$ then from Theorem 3.1

$$(4.6) \qquad \pi_{ij}(H) = \pi_{\bar{a}}(H_{\bar{a}}^+) = F^{\text{even}}(0, 1, H_{\bar{a}}^+).$$

If $G_{\bar{a}}^+$ is the undirected graph obtained by ignoring the directing of the arcs of $H_{\bar{a}}^+$ then the number of undirected mod λ flows $F(\alpha, \lambda, G_{\bar{a}}^+)$, with a nonzero flow of α on $\bar{a}$ is independent of α and is given by $\lambda^{c(G)} \gamma_{ij}(G)$, where $c(G)$ is the cycle rank of G. It follows that

$$(4.7) \qquad \gamma_{ij}(G) = F(0, 1, G_{\bar{a}}^+),$$

which is directly analogous to (4.6).

Equation (4.1) may be generalised to give the expected number $S_i^{(m)}(p, H)$ of vertices which are m-connected to u

$$(4.8) \qquad S_i^{(m)}(p, H) = \sum_j C_{ij}^{(m)}(p, H),$$

where $C_{ij}^{(m)}(p, H)$ is the probability that vertex j is m-connected from i. For the asymptotics of $S_i^{(m)}(p, H)$ a generalisation of the d-weight is needed together with simplifying rules, see [2]. The critical probability $p_c^{(m)}$ for this function on an infinite graph gives the onset of the existence of infinite clusters of vertices which are m-connected to i.

4.2. Vicious walkers and polymer networks on a lattice. Fisher, in his Boltzmann medal award lecture, [18], considered the problem of m lock-step walkers who start from distinct points on the real line having even integer co-ordinates and at each time step simultaneously, but independently, move a unit distance to the left or right with equal probability. *Vicious walkers* shoot one another if they arrive at the same point. The problem Fisher addressed was the determination of the *survival probability* $P_S(t)$ that m vicious walkers all stay alive for at least t steps. He also considered the *reunion probability* $P_R(t)$ that they survive and end up at distance two apart.

The space-time trajectories of the walkers are paths on a directed square lattice the sites of which are the points of the plane having integer co-ordinates the sum of which is even. These paths may be pictured as the embeddings of directed polymer chains of length t no pair of which intersect [17].

To make the problem specific let us suppose that the ith walker has initial position $x_i = 2(i - 1)$, then

$$(4.9) \qquad P_S(t) = S_t(m)/2^{mt},$$

where $S_t(m)$ is the number of possible t-step paths of the m walkers which are nonintersecting and end anywhere. In polymer terminology this is the number of *star configurations*. To obtain the reunion probability further suppose that after t steps, $x_i = 2(i - 1) + 2q - t$, where q is the number of positive steps made by each walker, then

$$(4.10) \qquad P_R(t, q) = w_t(m, q)/2^{mt},$$

where $w_t(m, q)$ is the number of nonintersecting path configurations subject to the above initial and final conditions. Again in polymer terminology this is the number of *watermelon configurations*.

In [**10**], we formulated the reunion problem in terms of even flows on the directed square lattice. This is made possible by the following bijection between the paths of m-vicious walkers and flows.

Consider the configuration of the m walkers in the x, t-coordinate plane. The assumption of nonintersection means that if we translate the path of the i-th walker by $2(i - 1)$ in the negative x-direction we then have m paths all beginning at the point $(0, 0)$ and terminating at (k, t) where $k = 2q - t$. Moreover, the translated paths taken pairwise do not cross over although they might share common arcs or vertices. If a unit flow is attached to each walk, the above translation of paths then provides an essentially rooted and directed integer flow from root $(0, 0)$ to root (k, t) with a flow of m through the "root" vertices. Note that this is a $\mathbb{Z}$-flow in the sense that the Kirchhoff constraint at each vertex is zero over the *integers*. This is equivalent to considering an even mod λ flow, for odd $\lambda > 2m$, with root flow $2m$ obtained by attaching a flow of 2 to each walker. The odd mod λ Kirchhoff constraint for an even flow at a grid vertex with two edges oriented in and out forces a $\mathbb{Z}$-flow.

Conversely, a $2m$ rooted even flow of the type described above can be uniquely unfolded to produce m parallel walks with a flow of 2 for each walk.

It follows from Theorem 3.1 that these walk configurations are enumerated by a polynomial in m, see also [**10, 17**]. This is not apparent in the Fisher approach where m appears as the dimension of a matrix.

A direct argument which shows that the number of the above $\mathbb{Z}$-flows may be counted by a polynomial in m is as follows. We let $\mathcal{P}$ be the set of all possible walker paths from source $(x, t) = (0, 0)$ to sink (k, t). They can be given the structure of a partially ordered set where two given paths ϕ, ϕ' satisfy $\phi < \phi'$ if the path ϕ' is to the right of ϕ in the x, t-plane. The number of flows is then obtained by distributing the flow m among the paths in any totally ordered set of $\mathcal{P}$. If we denote the family of nonempty totally ordered subsets of $\mathcal{P}$ by Θ, the number of vicious walker configurations [**10**] is

$$(4.11) \qquad w_t(m, q) = \sum_{\theta \in \Theta} \frac{(m - |\theta| + 1)_{|\theta| - 1}}{(|\theta| - 1)!},$$

where we have used the Pochhammer symbol $(a)_k = a(a + 1)(a + 2) \dots (a + k - 1)$ for $k > 0$ and $(a)_0 = 1$. The RHS of (4.11) is clearly polynomial in m.

For $0 \le q \le \frac{1}{2} t$ we have the explicit product form of the polynomial,

$$(4.12) \qquad w_t(m, q) = \prod_{j=1}^{q} \frac{(m + j)_{t - 2j + 1}}{(j)_{t - 2j + 1}}.$$

This formula was conjectured in [**10**] and proved in [**17**]. An equivalent formula, valid for $0 \le q \le t$ but not explicitly of polynomial form in m, is

$$(4.13) \qquad w_t(m, q) = \prod_{i=0}^{m-1} \frac{(t - q + i + 1)_q}{(i + 1)_q} = \prod_{i=0}^{m-1} \left[\frac{i!}{(t + i + 1)_i} \binom{t + 2i}{q + i} \right].$$

A Gaussian approximation to the binomial coefficient in (4.13) gives the asymptotic form as $t \to \infty$

$$(4.14) \qquad w_t(m, q) \cong \left(\prod_{i=1}^{m-1} i! \right) \left[(2\pi)^{-1/2} 2^{m+t} t^{-m/2} \exp(-x_1^2 / 2t) \right]^m.$$

In [17] it was shown that equation (4.12) can be generalised to the case when the ith walker makes q_i positive steps and $q_i \geq q_{i-1}$. The number of t-step configurations is given by

$$(4.15) \quad w_t(m, q_1, q_2, \ldots, q_m)$$

$$= \prod_{1 \leq i < j \leq m} (q_j - q_i + j - i) \prod_{j=1}^{m} \frac{(t + m - j)!}{(q_j + j - 1)!(t - q_j + m - j)!}.$$

The number of m vicious walker configurations $S_t(m)$ which finish anywhere after t steps may be obtained by summing (4.15) over the values of q_i subject to $0 \leq q_1 \leq q_2 \leq \cdots \leq q_m \leq t$. In [10] it was conjectured that

$$(4.16) \qquad S_t(m) = \prod_{j=1}^{\lfloor (t+1)/2 \rfloor} \frac{(m + 2j - 1)_{2t-4j+3}}{(2j - 1)_{2t-4j+3}},$$

where $\lfloor \ \rfloor$ is the floor function. The asymptotic form of $S_t(m)$ as $t \to \infty$ is

$$(4.17) \qquad S_t(m) \cong \left(\frac{2^{e(m-e/2)/2}}{\pi^{e/4}} \prod_{k=1}^{e/2} (m - 2k)! \right) 2^{mt} t^{-m(m-1)/4},$$

where $e = 2\lfloor m/2 \rfloor$. This result was also obtained by Fisher [18] using a continuum approximation.

More recently it has been brought to our attention [26] that (4.16) is a special case of the Bender-Knuth conjecture [11] on plane partitions which was first proved by Gordon [19]. The first published proof was by Andrews [1]. For a more general discussion between plane partitions and parallel walks, see [31,32] and [20]. To see the connection with plane partitions for the above case of vicious walkers, number the steps of each walk from 1 to t and let $n_{i,j}$ be the number of the jth positive step of the $(m - i + 1)^{\text{th}}$ walk, counting from the end of the walk, n_{ij} is zero for $i > m$ or if $j > t$. Clearly $n_{ij} > n_{i,j+1}$ and the mutual avoidance condition is expressed by $n_{ij} \geq n_{i+1,j}$. Such a matrix subject to $\sum_{ij} n_{ij} = n$ is known by Bender and Knuth as an m-rowed strict plane partition of n with no part exceeding t. They conjectured [11, eq. 8] that the number of such partitions is the coefficient of z^n in the expansion of the generating function

$$(4.18) \qquad \prod_{i=1}^{t} \prod_{j=i}^{t} \frac{1 - z^{m+i+j-1}}{1 - z^{i+j-1}}.$$

The number of star polymer configurations is the number of different matrices satisfying all of the above conditions except $\sum_{ij} n_{ij} = n$ and is therefore obtained by taking the limit $z \to 1$ in (4.18) with the result

$$(4.19) \qquad S_t(m) = \prod_{i=1}^{t} \prod_{j=i}^{t} \frac{m + i + j - 1}{i + j - 1} = \prod_{i=1}^{t} \frac{(m + 2i - 1)_{t-i+1}}{(2i - 1)_{t-i+1}},$$

which may be rearranged to give (4.16).

The result corresponding to equation (4.13) for plane partitions is found in [28] where $w_t(m, q)$ counts tableaux, or plane partitions, of rectangular shape with bounded entries. The more general enumeration in equation (4.15) counts tableaux of a given arbitrary shape. This, in turn, can be interpreted in terms of dimension formula for the irreducible representation of $SL(n)$ given in [27].

4.3. Potts models. It is possible to extend the simple counting of proper colourings on a graph to more refined weighted enumerations. Enumerations in this broader class are often referred to as interaction models, see [13] for a general discussion. Here we consider a subclass of these models, originally introduced by Potts in [30], called *Potts models.*

Let $A(G)$ be the arc set of some arbitrarily chosen directing of the graph G. The standard Potts model partition function is defined by

$$(4.20) \qquad Z_\lambda(G, u, v) = \sum_c \prod_{a \in A(G)} w\big(\delta c(a)\big),$$

where the sum is over all vertex colourings and the arc weight function w is given by

$$(4.21) \qquad w(\alpha) = \begin{cases} u & \text{if } \alpha \neq 0 \mod \lambda, \\ v & \text{if } \alpha = 0 \mod \lambda. \end{cases}$$

Here u, $v \in \mathbb{C}$, or more generally these weights can take values in a given ring. The partition function so defined is independent of the chosen directing and we call this the undirected Potts model.

The origin of these models is in physics where macroscopic properties of many particle structures, e.g. magnetization of materials, need to be explained in terms of local interactions between neighbouring atoms. In these physical applications the summation $Z_\lambda(G)$ is called the *partition function* and determines the thermodynamic properties of the system being modelled.

Partitioning the colourings in (4.20) according to the edge subset E' on which the colour difference is zero gives

$$(4.22) \qquad Z_\lambda(G, u, v) = \sum_{E' \subseteq E} P_\lambda^+(G/G') v^{|E'|} u^{|E \setminus E'|},$$

which evaluates to the chromatic polynomial of G when we make the substitutions $u = 1$ and $v = 0$. This result together with (1.3) shows that $Z_\lambda(G, u, v)$ is a polynomial in λ of degree at most $r(G) + 1$ as well as being polynomial in the variables u and v.

In this section we consider, for odd λ, chiral Potts models which have partition functions which do depend on the chosen directing of G. These partition functions can be expressed in terms of the directed polynomials discussed in this survey, see [4]. We also present here, for the first time, 'reversal-deletion-contraction' rules for chiral models which generalise those previously obtained for the directed chromatic polynomials in [5]. Finally, the partition functions described here will be used to develop knot invariants in the next section.

4.3.1. *Polynomial form of the λ-state chiral Potts model partition function.* The arc weights for the chiral Potts model are given by

$$(4.23) \qquad w(\alpha) = \begin{cases} u & \text{if } \alpha \mod \lambda \in \mathcal{C}^e, \\ \bar{u} & \text{if } \alpha \mod \lambda \in \mathcal{C}^o, \\ v & \text{if } \alpha = 0 \mod \lambda, \end{cases}$$

where $\mathcal{C}^e \subset \mathcal{C}$ is the set of nonzero even integers and $\mathcal{C}^o \subset \mathcal{C}$ is the set of odd integers. For $H \in \mathcal{D}(G)$ define the partition function $\vec{Z}$ by

$$(4.24) \qquad \vec{Z}_\lambda(H, u, \bar{u}, v) = \sum_c \prod_{a \in A(H)} w\big(\delta c(a)\big),$$

which reduces to the standard undirected Potts model partition function when $\bar{u} = u$ but it will depend on the chosen directing H of G if $\bar{u} \neq u$.

The formula for the chiral Potts model given in equation (4.24) has an explicit polynomial form in the variables u, $\bar{u}$, v. To show that the partition function $\vec{Z}$ also has polynomial dependence on λ our approach is to expand $\vec{Z}$ with coefficients which are generated by enumerations of even λ colourings.

For $H, H' \in \mathcal{D}(G)$ define $\varepsilon_{H,H'}\colon E \to \{0,1\}$ by $\varepsilon_{H,H'}(e)$ is zero or one according as the orientations of e in H, H' are respectively the same or opposite.

Consider $\vec{Z}$ on the directed graph $H = (V, A)$ of $\mathcal{D}(G)$ and firstly suppose that $v = 0$ so that the defining sum (4.20) becomes a sum over proper colourings.

$$(4.25) \qquad \vec{Z}_\lambda(H, u, \bar{u}, 0) = \sum_{c \in \mathcal{C} \setminus \{0\}} \prod_{a \in A} w\big(\delta c(a)\big).$$

The proper colourings may be enumerated by considering only colourings with even colour difference on some directing $H' \in \mathcal{D}(G)$ and then summing over all such directings, thus

$$(4.26) \qquad \vec{Z}_\lambda(H, u, \bar{u}, 0) = \sum_{H' \in \mathcal{D}(G)} \sum_{c \in \mathcal{C}^e} \prod_{e \in E(G)} u^{1 - \varepsilon_{H,H'}(e)} \bar{u}^{\varepsilon_{H,H'}(e)}$$

$$(4.27) \qquad = \sum_{H' \in \mathcal{D}(G)} u^{|A|} \left(\frac{\bar{u}}{u}\right)^{\rho(H,H')} P_\lambda^{\text{even}+}(H'),$$

where $\rho(H, H')$ is the number of arcs which must be reversed to obtain H' from H and $P_\lambda^{\text{even}+}(H')$ is the number of proper even colourings of H'. Together with Theorem 2.1 this shows that the values of $\vec{Z}_\lambda(H, u, \bar{u}, 0)$ for odd λ may be found by evaluating a polynomial of degree at most $r(G) + 1$ which is also a property of $\vec{Z}_\lambda(H, u, \bar{u}, v)$ since

$$(4.28) \qquad \vec{Z}_\lambda(H, u, \bar{u}, v) = \sum_{A' \subseteq A} v^{|A'|} \vec{Z}_\lambda(H/H', u, \bar{u}, 0).$$

The function $\vec{Z}_\lambda(H, u, \bar{u}, v)$ may also be written as a sum over directings of G. Since the summand in (4.28) is independent of the directing of the arcs in A' and there are $2^{|A'|}$ such directings, using (4.27) we find

$$(4.29) \quad \vec{Z}_\lambda(H, u, \bar{u}, v)$$

$$= \sum_{H' \in \mathcal{D}(G)} \sum_{A'' \subseteq A'} \left(\frac{v}{2}\right)^{|A''|} u^{|A' \setminus A''|} \left(\frac{\bar{u}}{u}\right)^{\rho(H,H',H'')} P_\lambda^{\text{even}+}(H'/H''),$$

where $\rho(H, H', H'')$ is the number of edges of H/H'' which must be reversed to obtain H'/H''.

4.3.2. *The reversal-deletion-contraction rule for λ-state chiral Potts models.* The RDC rule with respect to an arc $\bar{a} \in A(H)$ takes the form

$$(4.30) \quad \vec{Z}_\lambda(H, u, \bar{u}, v) + \vec{Z}_\lambda(H_{\bar{a}}^\rho, u, \bar{u}, v)$$
$$= (u + \bar{u})\vec{Z}_\lambda(H_{\bar{a}}^\delta, u, \bar{u}, v) + (2v - u - \bar{u})\vec{Z}_\lambda(H_{\bar{a}}^\gamma, u, \bar{u}, v)$$

where $H_{\bar{a}}^\rho$, $H_{\bar{a}}^\delta$ and $H_{\bar{a}}^\gamma$ are the directed graphs obtained from H by respectively reversing, deleting, contracting the arc $\bar{a} \in A$. Let the arc $\bar{a} = [i, j]$, then the rule can be shown using the following observations.

$$\vec{Z}_\lambda(H, u, \bar{u}, v) = \sum_{c_i = c_j} \prod + \sum_{c_j - c_i \in \mathcal{C}^e} \prod + \sum_{c_j - c_i \in \mathcal{C}^o} \prod$$
$$= v \sum_{c_i = c_j} \prod{}' + u \sum_{c_j - c_i \in \mathcal{C}^e} \prod{}' + \bar{u} \sum_{c_j - c_i \in \mathcal{C}^o} \prod{}',$$

where $\prod$ abbreviates the usual product of weights over all arcs and where $\prod'$ is the product over the arcs of the graph H' obtained by deleting the arc $\bar{a}$. Similarly,

$$\vec{Z}_\lambda(H_{\bar{a}}^\rho, u, \bar{u}, v) = \sum_{c_i = c_j} \prod + \sum_{c_j - c_i \in \mathcal{C}^e} \prod + \sum_{c_j - c_i \in \mathcal{C}^o} \prod$$
$$= v \sum_{c_i = c_j} \prod{}' + \bar{u} \sum_{c_j - c_i \in \mathcal{C}^e} \prod{}' + u \sum_{c_j - c_i \in \mathcal{C}^o} \prod{}'.$$

Thus

$$(4.31) \quad \vec{Z}_\lambda(H, u, \bar{u}, v) + \vec{Z}_\lambda(H_{\bar{a}}^\rho, u, \bar{u}, v)$$
$$= 2v \sum_{c_i = c_j} \prod{}' + (u + \bar{u})\left[\sum_{c_j - c_i \in \mathcal{C}^e} \prod{}' + \sum_{c_j - c_i \in \mathcal{C}^o} \prod{}' \right].$$

Finally,

$$\sum_{c_j - c_i \in \mathcal{C}^e} \prod{}' + \sum_{c_j - c_i \in \mathcal{C}^o} \prod{}'$$
$$= \left[\sum_{c_j - c_i \in \mathcal{C}^e} \prod{}' + \sum_{c_j - c_i \in \mathcal{C}^o} \prod{}' + \sum_{c_j - c_i = 0} \prod{}' \right] - \sum_{c_j - c_i = 0} \prod{}'$$
$$= \vec{Z}_\lambda(H_{\bar{a}}^\delta, u, \bar{u}, v) - \vec{Z}_\lambda(H_{\bar{a}}^\gamma, u, \bar{u}, v).$$

Note that the number of proper even- and odd- colourings can be evaluated from $\vec{Z}_\lambda(H, u, \bar{u}, v)$.

LEMMA 4.1. *Let $H = (V, A)$ be a planar directed graph, then the chiral partition function $\vec{Z}_\lambda(H, u, \bar{u}, v)$ satisfies*

$$(4.32) \quad \vec{Z}_\lambda(H, 1, 0, 0) = P_\lambda^{\mathrm{even}+}(H), \quad \vec{Z}_\lambda(H, 1, 0, 1) = P_\lambda^{\mathrm{even}}(H),$$

the proper and improper even colour difference chromatic polynomials respectively, and

$$(4.33) \quad \vec{Z}_\lambda(H, 0, 1, 0) = P_\lambda^{\mathrm{odd}+}(H), \quad \vec{Z}_\lambda(H, 0, 1, 1) = P_\lambda^{\mathrm{odd}}(H),$$

the proper and improper odd colour difference chromatic polynomials respectively.

Clearly, the RDC rule for the chiral Potts partition function generalises the deletion-contraction rule for the partition function defined in (4.20), which can be obtained by identifying $G = H = H^\rho$ and $u = \bar{u}$, to give with $Z_\lambda(G, u, v) = \vec{Z}_\lambda(H, u, u, v)$

$$(4.34) \qquad Z_\lambda(G, u, v) = u Z_\lambda(G_a^\delta, u, v) + (v - u) Z_\lambda(G_a^\gamma, u, v).$$

The partition functions given by equations (4.20), (4.24) can be extended to correspond to the rooted case of the colouring polynomials [4]. Essentially, a particular arc $\bar{a}$ of H is distinguished and the colour difference $\delta c(\bar{a}) = \beta$ is prescribed. The RDC rule for proper even difference potentials used for the inductive proof of Theorem 2.1 is derived from (4.30) with the substitutions $u = 1$, $\bar{u} = v = 0$ and $u = v = 1$, $\bar{u} = 0$ to give respectively

$$(4.35) \qquad P_\lambda^{\text{even}+}(H) + P_\lambda^{\text{even}+}(H_{\bar{a}}^\rho) = P_\lambda^{\text{even}+}(H_{\bar{a}}^\delta) - P_\lambda^{\text{even}+}(H_{\bar{a}}^\gamma),$$

and

$$(4.36) \qquad P_\lambda^{\text{even}}(H) + P_\lambda^{\text{even}}(H_{\bar{a}}^\rho) = P_\lambda^{\text{even}}(H_{\bar{a}}^\delta) + P_\lambda^{\text{even}}(H_{\bar{a}}^\gamma).$$

Finally, consider the correlation function

$$(4.37) \qquad U^{\text{even}}(\beta, \lambda, H, u) = \frac{\vec{Z}_\lambda^{12}(H, u, 0, 1)}{\vec{Z}_\lambda(H, u, 0, 1)},$$

where $\vec{Z}_\lambda^{12}(.)$ is defined by (4.24) with the sum restricted to states for which the colour difference between vertices 1 and 2 has the fixed value β. Let $H_{\bar{a}}^+$ be the graph obtained from H by adding the arc $\bar{a} = \{1, 2\}$. Then, [4,8], using Theorem 2.1,

$$(4.38) \qquad U^{\text{even}}(0, 1, H, u) = \sum_{A' \subseteq A} \chi_{\bar{a}}(H_{\bar{a}}^+/H')(1-u)^{|A'|} u^{|A \backslash A'|},$$

and so (cf. eq. (4.1)) we have

$$(4.39) \qquad U^{\text{even}}(0, 1, H, u) = \bar{C}_{12}(1 - u, H),$$

where $\bar{C}_{12}(p, H)$ is the probability of no open path from 2 to 1 in the dual directed percolation model of Dhar *et al.* [15] for which there is probability p of an arc being open in both directions and probability $1 - p$ of being open only in the direction of the arc.

Very recent progress by Tsuchiya and Katori in the further use of chiral Potts models in walk problems and directed percolation is reported in [34].

4.4. Invariants for knots. One of the standard ways of representing a knot K is to project its 3-dimensional form onto a planar knot diagram D_K which records the over and under- crossings. As we see below a graph can be associated with the knot diagram and a partition function can be attached to the graph. Clearly, there are many knot diagrams and therefore different partition functions, that can be produced depending on the different projections of K. The different knot diagrams of K are related by the 'Reidemeister' rules. Thus an invariant for the knot K can be found if the various weights can be chosen so that the resulting partition function remains identical over all knot diagrams of the knot K. This means that the partition function has to be left unchanged under the effect of each of the Reidemeister rules on graphs. Such invariants can be found by this approach and one of the simplest is the bracket polynomial.

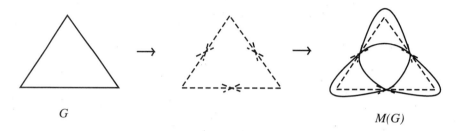

G $M(G)$

FIGURE 1. The construction of the knot universe $M(G)$ from a graph G.

4.4.1. *Bracket polynomial.* The following construction is developed by Kauffman in [**22**]. There are minor modifications to reflect the notation developed earlier in this paper. Given a planar graph G we can associate a knot universe $M(G)$ by the method of introducing a cross at the mid point of each edge $e \in E$ and then joining adjacent arms of the crosses.

Now consider $K(G)$, one of the two knots obtained by introducing alternating crossings to the planar universe $M(G)$, and convert it to the knot diagram D_K. Note that one is a reflection of the other. Thus $K(G)$ is a knot associated with the graph G. We define the bracket polynomial $\{K\}$ of a knot K to satisfy:

1. $\{ \overset{\frown}{\diagdown} \} = u \{ \rangle \langle \} + (v - u)\lambda^{-1/2}\{ \smile \}$;
2. $\{ \bigcirc \cup K \} = \lambda^{1/2}\{K\}$;
3. $\{ \bigcirc \} = \lambda^{1/2}$.

These properties should be compared with those of the following partition function on the graph $G = (V, E)$ with λ-colourings denoted by $c : V \to \mathbb{Z}_\lambda$. Consider a new partition function modified from (4.20)

$$(4.40) \qquad \widehat{Z}_\lambda(G, u, v) = \lambda^{-\nu/2} Z_\lambda(G, u, v).$$

The standard Potts partition function has weights given by

$$(4.41) \qquad w_e(c_i, c_j) = u + (v - u)\delta_{c_i, c_j},$$

where $e = [i, j] \in E$ and $\delta_{..}$ is the delta-function. This gives only two values, u and v. The deletion-contraction rule for $\widehat{Z}$ (cf. (4.34)) is given by

$$(4.42) \qquad \widehat{Z}_\lambda(G, u, v) = u\widehat{Z}_\lambda(G \backslash e, u, v) + (v - u)\lambda^{-1/2}\widehat{Z}_\lambda(G/e, u, v),$$

where $G \backslash e$ and G/e denote respectively the graph G with the edge e deleted and contracted. This is rule 1 for the Kauffman bracket. Rules 2 and 3 for the partition function, correspond to

1. the disjoint union of a single vertex and a graph G is given by

$$(4.43) \qquad \widehat{Z}_\lambda(\bullet \cup G, u, v) = \lambda^{1/2}\widehat{Z}_\lambda(G, u, v),$$

   and,
2. for a single vertex

$$(4.44) \qquad \widehat{Z}_\lambda(\bullet, u, v) = \lambda^{1/2}.$$

It can now be seen that the multiplicative factor in the definition of $\widehat{Z}_\lambda(G, u, v)$ is used the ensure the value $\lambda^{1/2}$ for the trivial graph of a single vertex.

FIGURE 2. A knot K is *reduced* if its diagram D_K does not divide into subsets D_1 and D_2 connected by a two-strand bridge.

PROPOSITION 4.2. [**22**] *Let $G = (V, E)$ be a planar graph and let $K(G)$ be an associated alternating knot, then*

$$(4.45) \qquad \left\{K(G)\right\} = \lambda^{-\nu/2} Z_\lambda(G, u, v).$$

Hence the knot bracket $\left\{K(G)\right\}$ can be written in the Potts formalism by using (4.20) to obtain

$$(4.46) \qquad \left\{K(G)\right\} = \lambda^{-\nu/2} \sum_c \prod_{e=[i,j]\in E} (u + (v - u)\delta_{c_i,c_j}).$$

Furthermore, if we simplify the coefficients of the Kauffman bracket and introduce $[K] = [K](A, B, d)$, defined by

1. $[\curvearrowright] = A[\smile] + B[)($;
2. $[\bigcirc \cup K] = d[K]$;
3. $[\bigcirc] = d$,

is an invariant, [**24**], for *reduced* alternating knots for all A, B and d. Thus we have a knot invariant $[K](A, B, d) = \left\{K(G)\right\}$ which is given explicitly in terms of the Potts model partition function $Z_\lambda(H, u, v)$ by choosing

$$(4.47) \qquad \lambda = d^2, \quad u = B, \quad v = Ad + B.$$

The bracket $[\cdot]$ is enumerated in the integer ring $\mathbb{Z}[A, B, d]$. This follows from the inductive nature of the definition of the bracket where rule 1 gives the bracket of a knot with n-crossings in terms of the bracket for knots with fewer crossings.

The reason that the bracket $[\cdot]$ is an invariant of reduced alternating knots *for all A, B, d* is that there is an alternative equivalence for alternating knots. The Reidemeister rules RII and RIII can be replaced by imposing invariance under 'flyping moves', [**24**]. The property was originally conjectured by Tait in [**33**] and was recently proven in [**29**].

TAIT'S THIRD CONJECTURE. Any two reduced knot diagrams D_K and D'_K of an alternating knot K are equivalent by performing a finite number of flypes.

A *flype* is a rotation of a $(2,2)$-tangle in a knot as indicated in Fig. 3. The flype leaves the class of reduced alternating knots invariant and hence allows their equivalence to be addressed within the class of alternating knots.

The important feature of a flype is that the associated Potts models remain unchanged for all choices of the triple λ, u, v and thus the knot bracket $[K](A, B, d)$ is a *regular isotopy* knot invariant of an alternating knot K for all choices of A, B and d. For nonalternating knots, the invariance requires further constraints on A, B and d to ensure invariance under RII and RIII.

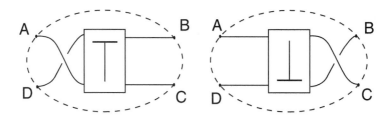

FIGURE 3. The flype construction on a $(2,2)$-tangle of a knot K. The subknot denoted by 'T' is rotated by a half-turn, keeping the points A, B, C and D fixed, to form '⊥' and the rest of the knot outside the elliptical boundary is left unchanged.

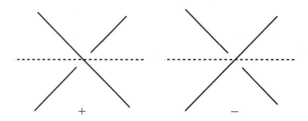

FIGURE 4. The signing of knot crossings and edges (dashed) in a knot diagram.

4.4.2. *Dichromatic polynomial.* It should also be noted that if we take $u = 1$ and introduce $\bar{v} = (v - 1)$ in the deletion-contraction rule (4.34) for Z , we obtain

$$(4.48) \qquad Z_\lambda(G, \bar{v}) = Z_\lambda(G_a^\delta, \bar{v}) + \bar{v}Z_\lambda(G_a^\gamma, \bar{v}).$$

The partition function $Z = \lambda^{\nu/2}\widehat{Z}$ is the *dichromatic polynomial* [**25**] which satisfies $Z_\lambda(\bullet, \bar{v}) = \lambda$ for a single vertex and $Z_\lambda(\bullet \cup G, \bar{v}) = \lambda Z_\lambda(G, \bar{v})$. The standard chromatic polynomial can be recovered from the dichromatic polynomial by choosing $\bar{v} = -1$, i.e. $v = 0$ so that

$$(4.49) \qquad Z_\lambda(G, -1) = \lambda^{\nu/2}\widehat{Z}_\lambda(G, -1) = \lambda^{\nu/2}\big\{K(G)\big\}\big|_{u=1,v=0}.$$

We conclude that the chromatic polynomial for the graph G can be found by evaluating the knot bracket of the associated knot diagram of $K(G)$.

4.4.3. *Signed graphs.* The Potts model can be extended to a signed graph G^s of G where $s\colon E \to \{-, +\}$ denotes an edge-signing of G. This enables us to distinguish different types of crossing (see Fig. 4). Note that in the case of alternating knots, all crossings are of the same type, i.e. either all $-$ or all $+$.

With the above notation define

$$(4.50) \qquad \widehat{Z}_\lambda(G^s, \mathbf{u}, \mathbf{v}) = \lambda^{-\nu/2}\sum_c \prod_{e=[i,j]\in E} w_e^{s(e)}(c_i, c_j),$$

where $\mathbf{u} = (u^-, u^+)$, $\mathbf{v} = (v^-, v^+)$ and

$$(4.51) \qquad w_e^\star(c_i, c_j) = u^\star + (v^\star - u^\star)\delta_{c_i, c_j},$$

for $\star = \pm$. We are now able to associate weights dependent on signing of the edges. The signing also gives the partition function a 'deletion-contraction' rule for the edge $e^\star$ as

$$(4.52) \qquad \widehat{Z}_\lambda(G^s, \mathbf{u}, \mathbf{v}) = u^\star \widehat{Z}_\lambda(G^s \backslash e^\star, \mathbf{u}, \mathbf{v}) + (v^\star - u^\star)\lambda^{-1/2}\widehat{Z}_\lambda(G^s/e^\star, \mathbf{u}, \mathbf{v}).$$

To complete the analogous three rules to those of the knot bracket, the partition function of the disjoint union of a single vertex and a graph G is given by

$$(4.53) \qquad\qquad \widehat{Z}_\lambda(\bullet \cup G^s, \mathbf{u}, \mathbf{v}) = \lambda^{1/2}\widehat{Z}_\lambda(G^s, \mathbf{u}, \mathbf{v}),$$

and for a single vertex

$$(4.54) \qquad\qquad\qquad \widehat{Z}_\lambda(\bullet, \mathbf{u}, \mathbf{v}) = \lambda^{1/2}.$$

Comparison of coefficients between the reduction formulae for $[K]$ and for $\widehat{Z}_\lambda(G^s, \mathbf{u}, \mathbf{v})$ gives

LEMMA 4.3. [24] *For a knot K, the bracket $[K](A, B, d)$ and the partition function $\widehat{Z}_\lambda(G_K^s, \mathbf{u}, \mathbf{v})$ are identical when the equations*

$$(4.55) \qquad u^- = (v^+ - u^+)\lambda^{-1/2} = A, \quad u^+ = (v^- - u^-)\lambda^{-1/2} = B, \quad \lambda^{1/2} = d,$$

are satisfied.

REMARK. To ensure that $[K]$ is a regular knot invariant we require the further conditions, namely $AB = 1$ and $A^2 + B^2 + d = 0$, cf. equation (4.47) and [24, 25]. This, of course, imposes further constraints on the weights $u^\pm, v^\pm$ of the Potts model. The more typical approach is to produce an invariant which reduces the number of variables in the partition function. A classical example of this is the Jones polynomial which is obtained as a special case of the Kaufmann bracket, and hence of the partition function, for oriented knots, cf. Lemma 4.3 and [24].

For the case of an alternating knot K, we have observed that the signings are either all $+$ or all $-$. Thus, if required, we can use one of the two signings to replace the variables u, v with either the pair u^+, v^+ or the pair u^-, v^- respectively. Furthermore, the reduction rules for the bracket $[K]$ (respectively the Potts partition function $Z_\lambda(G_K^s, u, v)$) leave the crossing types(signed edges) all of one sign. Finally the flype does not change the type of crossings as the new knot is still alternating. In the following, the weights $w_e(c)$ attached to each edge $e = [i, j]$ can be either all $w_e^+(c_i, c_j)$ or all $w_e^-(c_i, c_j)$. We will denote this choice by $\star$ and assume therefore that the substitution $\star \equiv +$ or $\star \equiv -$ is made.

LEMMA 4.4. *Let the knots K and K' differ by a single flype, then the Potts partition functions $Z_\lambda(G_K^\star, u, v)$ and $Z_\lambda(G_{K'}^\star, u, v)$ are identical.*

REMARK. The flype construction does not change either the sign of crossings or their number and so the same Potts models arise from two equivalent alternating knots with the *same* number of crossings. We therefore conclude from the theorem that two alternating knots with the same number of crossings are not equivalent if they have different Potts models.

4.4.4. *Chiral partition functions and alternating knot invariants.* The idea from the previous section can now be extended to the set of directings $\mathcal{D}(G_K)$ of the undirected graph $G_K = (V, E)$. Typically, $H_K = (V, A) \in \mathcal{D}(G_K)$, where A is an arc set obtained by directing E. Furthermore, define $\mathbb{Z}_\lambda^e(\mathbb{Z}_\lambda^o) \subset \mathbb{Z}_\lambda$ to be the image of the even integers (odd integers) in the integer interval $[1, \lambda - 1]$ under the natural projection $p \colon \mathbb{Z} \to \mathbb{Z}_\lambda$. A chiral partition function is given in [4]. Here we discuss a normalised version given by

$$(4.56) \qquad \widehat{\widehat{Z}}_\lambda(H_K, \mathbf{u}, \bar{\mathbf{u}}, \mathbf{v}) = \lambda^{-\nu/2} \sum_c \prod_{a=[i,j]\in A} w_a(c).$$

For a given signing s of the directed graph H_K, we consider the partition function

$$(4.57) \qquad \widehat{\widehat{Z}}_\lambda(H_K^s, \mathbf{u}, \bar{\mathbf{u}}, \mathbf{v}) = \lambda^{-\nu/2} \sum_c \prod_{a=[i,j]\in A} w_a^{s(a)}(c),$$

where $\mathbf{u} = (u^-, u^+)$, $\bar{\mathbf{u}} = (\bar{u}^-, \bar{u}^+)$, $\mathbf{v} = (v^-, v^+)$ and the weight $w_a^\star$, where the arc $a = [i, j]$ and $\star = \pm$, is defined by

$$(4.58) \qquad w_a^\star(c) = \begin{cases} v^\star & c_j - c_i = 0, \\ u^\star & c_j - c_i = \mathbb{Z}_\lambda^e, \\ \bar{u}^\star & c_j - c_i = \mathbb{Z}_\lambda^o. \end{cases}$$

Note that for an alternating knot the signing is the same for all crossings in the knot diagram of K.

LEMMA 4.5. *Let $G_K^\star$ and $G_{K'}^\star$ be as in Lemma 4.4, then given a directing H_K of G_K, there exists a unique directing $H_{K'}$ of $G_{K'}$ such that*

$$(4.59) \qquad \widehat{\widehat{Z}}_\lambda(H_K^\star, \mathbf{u}, \bar{\mathbf{u}}, \mathbf{v}) = \widehat{\widehat{Z}}_\lambda(H_{K'}^\star, \mathbf{u}, \bar{\mathbf{u}}, \mathbf{v}).$$

The existence of a bijection of the directed graphs $H_{K'}$ and H_K can be used to obtain an invariant for knots by considering the totality of directings of G_K and $G_{K'}$. Given an alternating knot K, define

$$\mathbf{D}_K = \{\widehat{\widehat{Z}}_\lambda(H_K, \mathbf{u}, \bar{\mathbf{u}}, \mathbf{v}) | H_K \in \mathcal{D}(G_K)\},$$

to denote the class of chiral partition functions obtained from all directings of G_K.

PROPOSITION 4.6. *If K and K' are equivalent alternating knots then $\mathbf{D}_K = \mathbf{D}_{K'}$.*

Finally, we observe that the standard Potts partition function, and therefore the Jones polynomial, can be obtained from the chiral version.

PROPOSITION 4.7. *For a given knot K, the Potts partition function can be obtained from the chiral Potts model of a unique directing H_K of G_K by the relation*

$$(4.60) \qquad \widehat{Z}_\lambda(G_K, \mathbf{u}, \mathbf{v}) = \widehat{\widehat{Z}}_\lambda(H_K, \mathbf{u}, \mathbf{u}, \mathbf{v}).$$

References

1. G. Andrews, *Plane partitions. II. The equivalence of the Bender-Knuth and MacMahon conjectures*, Pacific J. Math. **72** (1977), 283–291.
2. D. K. Arrowsmith, *Percolation theory on multi-rooted directed graphs*, J. Math. Phys. **20** (1979), 101–3.
3. D. K. Arrowsmith and J. W. Essam, *Percolation theory on directed graphs*, J. Math. Phys. **18** (1977), 235–238.

4. _____ , *Extension of the Kasteleyn-Fortuin formulas to directed percolation*, Phys. Rev. Lett. **65** (1990), 3068–3071.
5. _____ , *Chromatic and flow polynomials for directed graphs*, J. Combin. Theory Ser. B **62** (1994), no. 2, 349–362.
6. _____ , *Reciprocity and polynomial properties for even flows and potentials on directed graphs*, Combin. Probab. Comput. **3** (1994) 1–11.
7. _____ , *Percolation theory on directed graphs*, J. Math. Phys. **18** (1977), no. 2, 235–238.
8. _____ , *Restricted colourings and flows on graphs and directed percolation*, Trends in Stat. Phys. **1** (1994), 143–152.
9. _____ , *Chiral Potts models and knot invariants*, preprint.
10. D. K. Arrowsmith, J. W. Essam, and P. Mason, *Vicious walkers, flows and directed percolation*, Current Problems in Statistical Mechanics (Washington, 1991), Phys. A **177** (1991), no. 1–3, pp. 267–272.
11. E. A. Bender and D. E. Knuth, *Enumeration of plane partitions*, J..Combinatorial Theory Ser. A **13** (1972), 40–54.
12. N. L. Biggs, *On the duality of interaction models*, Math. Proc. Cambridge Philos. Soc. **80** (1976), 429–436.
13. _____ , *Interaction Models*, Course given at Royal Holloway College (Univ. London, 1976), London Math. Soc. Lecture Note Ser., no. 30, Cambridge Univ. Press, Cambridge-New York-Melbourne, 1977.
14. S. R. Broadbent and J. M. Hammersley, *Percolation processes. I. Crystals and mazes*, Proc. Cambridge Philos. Soc. **53** (1957), 629–641.
15. D. Dhar, M. Barma, and M. K. Phani, *Duality transformations for two-dimensional directed percolation and resistance problems*, Phys. Rev. Lett. **47** (1981), 1238–1240.
16. J. W. Essam, *Connectedness and connectivity in percolation theory*, Ann. Discrete Math. **33** (1987), 41–57.
17. J. W. Essam and A. J. Guttmann, *Vicious walkers and directed polymer networks in general dimensions*, Phys. Rev. E **52** (1995), 5849–5862.
18. M. E. Fisher, *Walks, walls, wetting, and melting*, J. Stat. Phys. **34** (1984), 667–729.
19. B. Gordon, *A proof of the Bender-Knuth conjecture*, Pacific J. Math. **108** (1983), 99–113.
20. A. J. Guttmann, A. L. Owczarek, and X. G. Viennot, *Vicious walkers and Young tableaux I. Without walls*, J. Phys. A **31** (1998), 3123–8135.
21. P. W. Kasteleyn and C. M. Fortuin, *On the random-cluster model. I. Introduction and relation to other models*, Physica **57** (1972), 536–564.
22. L. H. Kauffman, *On knots*, Annals of Math. Stud., vol 115, Princeton Univ. Press, Princeton, NY, 1987.
23. _____ , *An invariant of regular isotopy*, Trans. Amer. Math. Soc. **318** (1990), no. 2, 417–471.
24. _____ , *Statistical mechanics and the Jones polynomial*, Braids (Santa Cruz, 1986), Contemp. Math., vol. 78, Amer. Math. Soc., Providence, RI, 1988, pp. 263–297.
25. _____ , *A Tutte polynomial for signed graphs*, Combinatorics and Complexity (Chicago, 1987), Discrete Appl. Math. **25** (1989), no. 1–2, 105–127.
26. C. Krattenhaler, private communication.
27. D. E. Littlewood, *The theory of group characters and matrix representations of groups*, Oxford Univ. Press, New York, 1940.
28. P. A. MacMahon, *Combinatory analysis*, Two volumes (bound as one), Chelsea Publishing Co., New York, 1960.
29. W. Menasco and M. Thistlethwaite, *The classification of alternating links*, Ann. of Math. (2) **138** (1993), 113–171.
30. R. B. Potts, *Some general order-disorder transformations*, Math. Proc. Cambridge Philos. Soc. **48** (1951), 106–110.
31. D. Stanton and D. White, *Constructive Combinatorics*, Springer-Verlag, New York, 1986.
32. J. R. Stembridge, *Nonintersecting paths, pfaffians and plane partitions*, Adv. Math. **83** (1990), 96–131.
33. P. G. Tait, *On knots. I; II; III*, Scientific papers, vol. 1, Cambridge Univ. Press, London, 1898, pp. 273–347.
34. T. Tsuchiya and M. Katori, *Chiral Potts models, friendly Walkers and directed percolation problem*, J. Phys. Soc. Japan **67** (1998), no. 5, 1655–1666.

35. W. Tutte, *A contribution to the theory of chromatic polynomials*, Canad. J. Math. **5** (1954), 80–91.
36. H. Whitney, *The coloring of graphs*, Ann. of Math. (2) **33** (1932), 688–718.

SCHOOL OF MATHEMATICAL SCIENCES, QUEEN MARY AND WESTFIELD COLLEGE, UNIVERSITY OF LONDON, MILE END ROAD, LONDON, U.K
E-mail address: D.K.Arrowsmith@qmw.ac.uk

DEPARTMENT OF MATHEMATICS, ROYAL HOLLOWAY AND BEDFORD NEW COLLEGE, UNIVERSITY OF LONDON, EGHAM, SURREY, U.K
E-mail address: j.essam@rhbnc.ac.uk

Centre de Recherches Mathématiques
CRM Proceedings and Lecture Notes
Volume **23**, 1999

Four-Coloring Six-Regular Graphs on the Torus

Karen L. Collins and Joan P. Hutchinson

Dedicated to Herbert S. Wilf in honor of his 65th birthday, and to Phyllis Cassidy in honor of her retirement from Smith College

1. Introduction

A classic 3-color theorem attributed to Heawood [**10, 11**] (also observed by Kempe [**16**]) states that every *even triangulation* of the plane is 3-chromatic; by *even* we mean that every vertex has even degree. We are interested in the extent to which an analogous theorem carries over to even triangulations of the torus and to other surfaces of larger genus. In general, even triangulations of the torus, like K_7 and those containing K_4, are not 3-colorable, but it turns out that 4-coloring is very often possible.

It is known that the problem of determining 3-colorable graphs on the plane is an NP-complete problem (see [**8**]), and so 3-color theorems like those for triangle-free planar graphs and their generalizations have been highly prized; for surveys, see [**15, 21, 22**]. We concern ourselves instead with even triangulations, and we speculate that determining the chromatic number of such triangulations of higher-genus surfaces, including the torus, will yield good theorems. In general the chromatic number of even triangulations can grow with the genus, since for infinitely many values of n, the complete graph K_n can triangulate a surface (this follows from the genus formula for K_n [**19**]). The *edge-width* of such embeddings (i.e., the length of the shortest noncontractible cycle) is always three. Therefore we are interested in even triangulations with edge-width at least four, or even triangulations where the edge-width is an increasing function of the genus of the embedding surface.

This idea of looking at graphs embedded with large edge-width has been a fruitful one, especially for determining the chromatic properties of such graphs, see [**24, 26, 27**]. For a general graph embedded on an orientable surface of genus $g > 0$, $\lfloor (7 + \sqrt{48g + 1})/2 \rfloor$ colors are necessary and sufficient [**10, 19**]; however, Thomassen [**26**] has shown that a graph embedded on an orientable surface of genus $g > 0$ can be 5-colored provided all noncontractible cycles have length at least $2^{(14g+5)}$. This result was first shown for the torus in [**3**]. Hence these large

1991 *Mathematics Subject Classification*. Primary: 05C15; Secondary: 05C10.

We would like to thank the following people for helping us with this work: M. O. Albertson, L. Carter, D. Fisher, S. Fisk, M. Hovey, L. Krompart, and C. Thomassen. We would also like to thank the referees for helpful corrections and comments.

This is the final version of the paper.

edge-width graphs do not differ greatly in chromatic number from planar graphs, which can be 4-colored [5, 20].

Another example of a planar k-coloring theorem with a companion $(k + 1)$-coloring theorem for the same class of graphs embedded on surfaces, but with large edge-width, is the following. Call graphs that can be embedded in a surface with each region bounded by an even number of edges *evenly embeddable*. It is well-known that a planar graph can be 2-colored if and only if it is evenly embeddable in the plane. Also, a graph that is evenly embeddable in an orientable surface of genus $g > 0$ can be $\lfloor (5 + \sqrt{32g - 7})/2 \rfloor$-colored [12], but if it is embedded with edge-width at least $2^{(3g+5)}$, then three colors suffice [13]. The advantage of large edge-width is that the graphs are embedded in a "locally planar" fashion [3]. Thus, we conjecture the following.

CONJECTURE 1.1. *Every graph, embedded on an orientable surface of positive genus as an even triangulation and with edge-width sufficiently large, can be 4-colored.*

In this paper we focus on the torus and on ostensibly the most difficult case when the graph is a 6-regular triangulation; in [28] Thomassen asks for a characterization of the non-4-colorable, 6-regular graphs on the torus. A theorem of Altshuler [4] shows that every 6-regular toroidal graph can be represented as a 6-regular shifted rectangular grid on the torus; see Fig. 2. A shift of one is the same as an unshifted grid. This allows us to focus first on 6-regular grids and then to extend our colorings to the shifted grids. We show that:

THEOREM 1.2. *Let G be a graph embedded in a 6-regular $m \times n$ grid on the torus with a shift of i. If m, $n \geq 3$, then G can be 4-colored, with a finite number of exceptions. If $n = 1$ or 2, and $m \geq 3$, then there is an n_i such that every such graph with at least n_i vertices can be 4-colored.*

Lemmas 3.1, 3.2, 3.4 and Theorem 3.3 handle the unshifted $m \times n$ grids for m, $n \geq 3$. Theorem 3.6 provides the 4-coloring for the shifted $m \times n$ grids, for $3 \leq n$, m with m, $n \neq 5$, omitting some exceptions. Theorem 3.7 enumerates these exceptions. In Theorem 3.8 we show that the $m \times 2$ shifted grids are all 4-colorable for m even, and for m odd can be represented either as a $2m \times 1$ grid or an $m' \times n'$ grid with $3 \leq n', m'$ and $2m = m'n'$. The case of $m \times 1$ shifted grids is handled in Theorem 3.9. Hence we obtain our main result. It is only in the $3 \leq n$ case that we are able to state a bound on edge-width that insures 4-colorability.

THEOREM 1.3. *If G is a 6-regular toroidal graph, embedded as an $m \times n$ grid of any size shift, with $3 \leq m$, n and with edge-width at least 6, then G can be 4-colored.*

By results of [1, 2, 28], 6-regular graphs on the torus with edge-width at least 4 can be 5-colored; see both [28] and below (see the $(m \times 1; i)$ grids in Section 3) for an infinite family of 5-chromatic, 6-regular toroidal graphs, all with edge-width three. There are also sporadic 5-chromatic, 6-regular toroidal triangulations of edge-width at most 5; infinite families of 5-chromatic toroidal triangulations with arbitrarily large edge-width are known as are such graphs with exactly two vertices of odd degree [7].

The previous coloring and embedding questions apply to nonorientable surfaces as well. The chromatic number of all nonorientable surfaces is known [18] as it is for evenly embedded graphs on these surfaces [**9, 12**]. It is worth noting that though Thomassen's 5-color theorem carries over to nonorientable surfaces, there are 4-chromatic graphs that evenly embed on the projective plane with all regions 4-sided and with arbitrarily long noncontractible cycles [**30**]. It has recently been shown [**14**] that there are 5-chromatic even triangulations of the projective plane with arbitrarily large edge-width.

2. Background Material

We consider graphs that do not contain loops, but may contain multiple edges; we follow the terminology of [**15, 29**]. A graph is said to be k-*colored* (respectively, k-*colorable*) if each vertex of the graph is (resp., can be) assigned one of k colors so that no two adjacent vertices receive the same color. A graph is k-*chromatic* if k is the least integer so that the graph is k-colorable.

A graph is *planar* if it can be drawn in the plane (or equivalently on the surface of the sphere) without edge crossings; a graph *embeds* on a surface if it can be drawn there without edge crossings. The complement of a graph embedded on a surface is a set of open regions; the maximal connected open regions are called *faces* of the embedding. An embedding is a *triangulation* if each face is bounded by exactly three edges. The embedding is called a 2-*cell embedding* if each of the faces is a simply connected region. Crucial tools in the study of embedded graphs are the following:

Euler's Formula. If a connected graph is embedded in the plane with v vertices, e edges, and f faces, then $v - e + f = 2$.

Euler-Poincaré Theorem. If a graph has a 2-cell embedding with v vertices, e edges, and f faces on a surface of genus $g \geq 0$, then $v - e + f = 2 - 2g$.

Using standard counting arguments one gets the following.

COROLLARY 2.1. *If a multigraph with v vertices and e edges is embedded on a surface of genus $g \geq 0$ with each face bounded by at least three edges, then $e \leq 3v + 6(g - 1)$ with equality if and only if the embedding is a triangulation. The average degree of the graph, $2e/v$, is at most $6 + 12(g - 1)/v$.*

Applying the corollary to the torus where $g = 1$, we have that in general $e \leq 3v$ and so the average degree of a toroidal graph is at most 6. The embedding is a triangulation if and only if $e = 3v$, and then the average degree equals 6. Thus a graph on the torus either contains a vertex of degree less than six, or else the graph is 6-regular and a triangulation.

We use a characterization of 6-regular toroidal graphs, due to Altshuler [**4**]; a similar approach can be found in [**25**]. Imagine a rectangular grid of m rows and n columns; we label this by giving each grid point the label (i, j) where i denotes its row and j its column. Then define a 6-*regular right-diagonal (unshifted) $m \times n$ grid* on the torus to be the graph given by the vertex set $\{(i, j) \mid 1 \leq i \leq m, 1 \leq j \leq n\}$ where the neighbors of (i, j) are $(i, j - 1)$, $(i - 1, j)$, $(i - 1, j + 1)$, $(i, j + 1)$, $(i + 1, j)$, $(i + 1, j - 1)$ and arithmetic in the first coordinate is modulo m and in the second coordinate modulo n; see Fig. 1. These grids give a family of 6-regular toroidal graphs, but not all of them. We call the set of vertices $C_h(i) := \{(i, 1), (i, 2), \ldots, (i, n)\}$ a horizontal cycle and the set of vertices $C_v(j) := \{(1, j), (2, j), \ldots, (m, j)\}$ a vertical cycle of the grid.

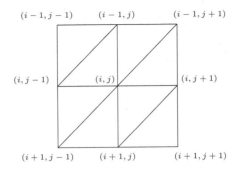

FIGURE 1. The neighbors of (i,j).

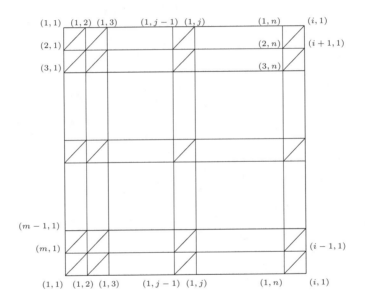

FIGURE 2. An $m \times n$ grid with a shift of i.

Now, we define a 6-*regular right-diagonal* $(m \times n; k)$ *grid* for some k, $1 \le k \le m$, to be the same as defined above except that there is a rotation before the vertices in the nth vertical cycle $C_v(n)$ are joined to those in the first, $C_v(1)$. Specifically the vertex of $C_v(n)$, (i,n), is now adjacent to vertices $(i+k-2,1)$ and $(i+k-1,1)$ as well as to $(i+1,n)$, $(i+1,n-1)$, $(i,n-1)$, $(i-1,n)$ for $i=1,2,\ldots,m$. In particular the first vertex of $C_v(n)$, $(1,n)$, is now adjacent to $(k-1,1)$ and $(k,1)$ in $C_v(1)$ and so a grid with shift 1 is the same as the unshifted grids, defined above.

THEOREM (ALTSHULER). [4] *Every 6-regular graph on the torus can be represented as a 6-regular right-diagonal $m \times n$ shifted grid on the torus.*

The parameters m and n may be found as follows: traverse a path in the embedded graph that, upon arriving at a vertex, continues "straight ahead," that is, takes the edge out that leaves two unused edges on either side of the path. As shown in [4] this process leads to a simple cycle, and we can label the vertices of this cycle $(1,1),(2,1),\ldots,(m,1)$. Next at vertex $(1,1)$, pick an unused edge on which to begin another straight-ahead path and continue until the original cycle is reached, say at vertex $(k, 1)$ after n vertices, labeled $(1, 1), (2, 1), \ldots, (n, 1)$. Then cutting the torus open along the first cycle and the second path produces an $(m \times n; k)$ grid. Note that there is choice in picking the unused edge at $(1, 1)$ so that a graph may be represented as a grid with more than one set of parameters.

3. Toroidal Results

In the following, read all first coordinates modulo m and second coordinates modulo n. We begin by showing how some colorings of a vertical (or horizontal) cycle in an $m \times n$ grid can be extended in the grid. Suppose that the vertical cycle $C_v(j)$ is vertex-colored by ϕ. We can extend ϕ to include the next vertical cycle $C_v(j + 1)$ by rotation of the coloring in two ways: Define $\phi_{-1}(i, j + 1) = \phi(i - 1, j)$ and $\phi_2(i, j + 1) = \phi(i + 2, j)$ for all i. (Descriptively, take the coloring on $C_v(j)$ and rotate it down one vertex, or up two vertices, and place that rotated coloring on the vertices of $C_v(j + 1)$.) These rotations need not be proper colorings; however, we shall show that if every successive triple of colors of ϕ on $C_v(j)$ is distinct, then ϕ_{-1} and ϕ_2 give proper coloring extensions. Similarly a coloring on the horizontal cycle $C_h(i)$ can be extended to the next horizontal cycle by one of two horizontal shifts.

LEMMA 3.1. *Let G be a 6-regular right-diagonal $m \times n$ (unshifted) grid on the torus where n is a positive integer, and $m \geq 3$. Suppose that the vertical cycle $C_v(j)$ of G is 3- or 4-colored so that the colors on each successive triple of vertices are distinct, i.e. for every k, $(k - 1, j)$, (k, j), $(k + 1, j)$ are all colored distinctly. Then ϕ_{-1} (respectively, ϕ_2) gives a proper coloring on $C_v(j) \cup C_v(j + 1)$ with the coloring on $C_v(j + 1)$ having the distinct triple property. In particular if $m = n$, applying ϕ_{-1} (resp., ϕ_2) successively to each of the other $n - 1$ columns of G gives a proper coloring. The same results hold with respect to horizontal cycles.*

PROOF. Without loss of generality, let ϕ be the coloring of $C_v(1)$, and consider the coloring on $C_v(2)$ resulting by applying ϕ_{-1}. Then a vertex $(i, 2)$ has neighbors $(i - 1, 2), (i + 1, 2), (i, 1)$ and $(i + 1, 1)$ in $C_v(1) \cup C_v(2)$. The vertices $(i - 1, 2), (i, 2)$ and $(i + 1, 2)$ form a distinctly colored triple since they received the colors, respectively of $(i - 2, 1)$, $(i - 1, 1)$, and $(i, 1)$. Thus ϕ_{-1} can next be applied to $C_v(3)$ and so on. After $m = n$ rotations the coloring will be back to that on $C_v(1)$ and the coloring on $C_v(n) \cup C_v(1)$ will be proper. The same proof works using ϕ_2 and with horizontal cycles and their horizontal rotations.   □

Now we demonstrate a variety of cases to show that if G is an unshifted graph with edge-width at least four, it is 4-colorable. Note that an unshifted $m \times 1$ grid is full of loops and so we don't color these. The unshifted $m \times 2$ grids are easily understood: in these the vertices of $C_v(1)$ and $C_v(2)$ form a complete graph on four vertices, and a 4-coloring of this complete graph extends to all vertices if and only if m is even. All $m \times 2$ grids are multigraphs, embedded with edge-width two.

LEMMA 3.2. *Let G be a 6-regular right-diagonal $m \times 3l$ grid on the torus, where l is a positive integer and $m \neq 5$. Then G is 4-colorable.*

PROOF. Write $m = 3t + 4s$ where $t \geq 0$ and $s = 0, 1, 2$. Color $C_v(1)$ by s successive repetitions of 1 2 3 4 followed by t repetitions of 1 2 3 ; let this be the coloring ϕ. Extend this coloring to $C_v(2)$ and $C_v(3)$ by applying ϕ_{-1} twice.

Since an application of ϕ_2 to $C_v(3)$ returns the coloring to that of $C_{v(1)}$, this gives a proper coloring of the $m \times 3$ grid by Lemma 3.1, and therefore the coloring of the first $m \times 3$ subgrid can be repeated l times to make a proper coloring of the $m \times 3l$ grid.   □

THEOREM 3.3. *Let G be a 6-regular right-diagonal $m \times n$ grid on the torus. Then if m, $n \geq 3$, G is 4-colorable.*

PROOF. The case when one or both of m and n is 5 is handled in the Lemma 3.4 below. We can also assume that n is not divisible by 3 by Lemma 3.2. If $n = m$, we need only prove that there exists a coloring of an n-cycle in which every set of successive triples is colored distinctly, and then use Lemma 3.1. Write $n = 3v + 4u$ where $v \geq 0$ and $u = 1, 2$. We can do this unless $n = 5$. Then coloring an n-cycle by u repetitions of $1\,2\,3\,4$ and v repetitions of $1\,2\,3$ will ensure that every successive triple is distinct.

Suppose that $n > m \geq 3$, and $n, m \neq 5$. Write $m = 3t + 4s$ where $t \geq 0$ and $s = 0, 1, 2$. Color the first four vertices in each horizontal cycle of the $m \times n$ grid as follows: put in t sets of the three rows

$$
\begin{array}{cccc}
1 & 2 & 3 & 4 \\
3 & 4 & 1 & 2 \\
2 & 3 & 4 & 1
\end{array}
$$

and s sets of the four rows

$$
\begin{array}{cccc}
1 & 2 & 3 & 4 \\
4 & 1 & 2 & 3 \\
3 & 4 & 1 & 2 \\
2 & 3 & 4 & 1
\end{array}
$$

One can check that this is a proper coloring of the $m \times 4$ grid. Further, the vertical cycle $C_v(4)$ has every successive triple colored distinctly; call this coloring ϕ. If $n \equiv 1 \pmod 3$, then apply ϕ_{-1} twice to get a coloring on $C_v(5)$ and $C_v(6)$. Color $C_v(7)$ the same as $C_v(4)$. The coloring on the block $C_v(5)$, $C_v(6)$, $C_v(7)$ is then a proper coloring of an $m \times 3$ grid; hence it can be repeated $(n - 7)/3$ times. This is still a proper coloring of the entire grid, since $C_v(n)$ is colored the same as $C_v(4)$.

If $n \equiv 2 \pmod 3$, repeat the first block of $m \times 4$ colors to color an $m \times 8$ block. Then let ϕ be the coloring on $C_v(8)$ and apply ϕ_{-1} twice to color $C_v(9)$ and $C_v(10)$, and color $C_v(11)$ the same as $C_v(8)$. Repeat the $C_v(9)$, $C_v(10)$, $C_v(11)$ block $(n - 11)/3$ times to finish coloring the grid.

Note that if $m > n \geq 3$ (and $n, m \neq 5$), then a k-coloring of the $n \times m$ grid, as given above, yields a k-coloring of the $m \times n$ grid by taking the transpose.   □

There still remains the case when one or both of m, n is 5; using the transpose argument, we can suppose that $m = 5$.

LEMMA 3.4. *Any $5 \times n$ grid is 4-colorable if $n \geq 3$.*

PROOF. The method of distinct successive triples will not work for a $5 \times n$ grid, because every non-adjacent pair of vertices in a 5-cycle is connected by a path of length 2, and in a 3- or 4- coloring of a 5-cycle, at least one color must be used twice. Thus the proof we have is to demonstrate a proper coloring of the $5 \times n$ grid for $n = 3, 4, 5, 6, 7, 9$, and then show how to construct a coloring of a larger grid from one of these. Here are proper colorings of the 5×3, 5×4, 5×5, 5×6, 5×7 and 5×9 grids.

```
1 2 3     1 3 2 4     1 4 3 2 4
3 1 2     2 4 1 3     2 1 4 1 3
4 3 1     3 2 4 1     3 2 3 2 1
1 4 2     1 3 2 4     1 4 1 3 4
2 3 4     2 4 1 3     2 3 2 1 3

1 3 4 1 2 4     1 3 2 1 3 2 4
2 1 2 4 1 3     2 4 3 2 4 1 3
3 4 1 3 2 4     3 2 4 1 3 4 1
1 3 2 4 1 2     1 3 2 4 2 3 2
2 4 1 2 4 3     2 4 1 3 4 1 3

     1 3 4 1 3 4 1 2 4
     2 1 3 2 1 3 4 1 3
     3 4 1 3 2 1 3 2 4
     1 3 2 4 3 2 4 1 2
     2 4 1 2 4 1 3 4 3
```

Each of the 5×4, 5×6, 5×7, and 5×9 grids begins with the column 1 2 3 1 2 and the set $\{4, 6, 7, 9\}$ is a complete residue system (mod 4). To color a $5 \times n$ grid for $n \neq 5$, select the value of 4, 6, 7, 9 that n is congruent to (mod 4), say a, and adjoin $(n - a)/4$ copies of the 5×4 grid to the $5 \times a$ grid. This will be a proper coloring. $\qquad\square$

Note that except for the 5×5 grid, the first horizontal cycle $C_h(1)$ of each grid formed this way has the distinct triple property.

COROLLARY 3.5. *If G is a 6-regular right-diagonal $m \times n$ grid on the torus with $m, n \geq 3$, then there is a 4-coloring of G with the first horizontal cycle $C_h(1)$ and the first vertical cycle $C_v(1)$ having distinct triples unless m or $n = 5$. If $m = 5$ (respectively, $n = 5$), there is a 4-coloring with the first horizontal (resp., vertical) cycle having distinct triples unless $m = n = 5$.*

PROOF. It is routine to check that the distinct triple property holds in all cases of the colorings in the proofs of Lemmas 3.1, 3.2, 3.4 and Theorem 3.3. $\qquad\square$

Notice that no 3-color theorem is possible in this setting since a $m \times n$ grid is 3-colorable if and only if 3 divides both m and n. The $m \times 3$ grids have edge-width three, and even if the edge-width is increased to 4, such a 3-color theorem is not possible since one can check that the $4 \times n$ grids are never 3-colorable.

THEOREM 3.6. *Let G be a 6-regular right-diagonal $(m \times n; i)$ grid on the torus for some i, $1 < i \leq m$. Then if $3 \leq m, n$, G can be 4-colored except possibly in the case when $m = 5$, or when $i = 2$ and $n = m$ or $m + 1$, or when $i = 3$ and $n = m$.*

PROOF. We color using rotations as before except that we require extra columns due to the presence of the i-shift. If $m, y \geq 3$ (where y will be specified later) and $m \neq 5$, we color the (unshifted) $m \times y$ grid as in Theorem 3.3. Then in G we use this coloring on $C_v(j)$, $j = 1, 2, \ldots, y + 1$ with the coloring on $C_v(y + 1)$ identical to that on $C_v(1)$. Note that by Corollary 3.5 the coloring on $C_v(y+1)$ has the distinct triples property. (If $y < m$, then this cycle is a horizontal cycle in the $y \times m$ grid.) Then we use the remaining columns of G to make a transition from the coloring ϕ on $C_v(y + 1)$ to the coloring on $C_v(1)$, shifted to have the ith vertex at the top. We do this by using ϕ_{-1}, applying it successively $m - i + 1$ times on the columns $C_v(y + 2)$, ..., $C_v(y + m - i + 2)$. At this point the coloring on the final column is identical to that on $C_v(1)$ except that the coloring is rotated to match that of $C_v(1)$ shifted to have its ith vertex on top. Thus we have a valid coloring of the $(m \times n; i)$ grid with $n = y + m - i + 1$ provided that $y \geq 3$. The inequality $y \geq 3$ fails only when $i = 2$ and $n = m$ or $m + 1$, or when $i = 3$ and $n = m$.   □

THEOREM 3.7. *Let G be a 6-regular right-diagonal $(m \times n; i)$ grid on the torus with $3 \leq m, n$. Then G can be 4-colored except in the cases of a $(3 \times 3; 2)$, or $(3 \times 3; 3)$ grid; a $(5 \times 3; 2)$, or $(5 \times 3; 3)$ grid; and a $(5 \times 5; 3)$, or $(5 \times 5; 4)$ grid.*

PROOF. Suppose $m \neq 5$ and $i = 2$ and $n = m$, or $i = 2$ and $n = m+1$, or $i = 3$ and $n = m$. We will show that G can be 4-colored. Together with Theorem 3.6 this proves the result in the cases where $m \neq 5$. To account for the i shift, we now use the coloring rotation ϕ_2. As in the proof of Theorem 3.6, color the $m \times y$ grid and use this coloring in G on $C_v(j)$, $j = 1, 2, \ldots, y + 1$ with the coloring on $C_v(y + 1)$ identical to that on $C_v(1)$. Call the latter coloring ϕ.

Then if $i = 3$ we apply ϕ_2 to obtain a coloring on $C_v(y+2)$ which is identical to that on $C_v(1)$ but with the third vertex shifted to the top. Thus we have a proper coloring of the $(m \times n; 3)$ grid with $n = y + 1$ which is valid provided $n - 1 \geq 3$. The only impossible case is when G is a $(3 \times 3; 3)$ grid.

If $i = 2$, we apply ϕ_2 to obtain a coloring on $C_v(y + 2)$ and then apply ϕ_{-1} to obtain a coloring of $C_v(y + 3)$ which is identical to that on $C_v(1)$ but with the second vertex shifted to the top. Thus we have a proper coloring of the $(m \times n; 2)$ grid with $n = y + 2$ which is valid provided $n - 2 \geq 3$. The only impossible cases are when G is a $(3 \times 3; 2)$ grid; a $(3 \times 4; 2)$ grid; or a $(4 \times 4; 2)$ grid. The third case can be 4-colored by alternately 2-coloring the vertical cycles. The second case can be 4-colored by

$$
\begin{array}{cccc}
1 & 3 & 4 & 3 \\
2 & 1 & 2 & 4 \\
3 & 4 & 1 & 2
\end{array}
$$

It is straightforward to check that the size of the largest independent set in a $(3 \times 3; 2)$ or a $(3 \times 3; 3)$ grid is 2, and hence that neither of these can be 4-colored.

Now suppose $m = 5$ so that we cannot rotate a coloring on vertical cycles. First we show that $(5 \times n; 2)$ grids are 4-colorable for $n \geq 5$. Note that, up to color permutation and rotation, there are two 4-colorings of the 5-cycle: 1, 2, 3, 1, 2 and 1, 2, 3, 4, 2. Then the $(5 \times 2; 2)$ grid is 4-colorable by starting with either of the following or a rotation of them:

$$
\begin{array}{cccc}
1 & 3 & \quad 1 & 3 \\
2 & 4 & \quad 2 & 1 \\
3 & 2 & \quad 3 & 2 \\
1 & 3 & \quad 4 & 1 \\
2 & 4 & \quad 2 & 4
\end{array}
$$

Thus a $(5 \times n; 2)$ grid can be 4-colored, by first coloring the (unshifted) $5 \times (n-2)$ grid and then regardless of the coloring of $C_v(1)$, adding on the desired remaining two columns. This is valid provided $n \geq 5$.

Similarly a $(5 \times 4; 3)$ grid is 4-colorable with either

$$
\begin{array}{cccc}
1 & 3 & 2 & 1 \\
2 & 4 & 3 & 4 \\
3 & 2 & 1 & 3 \\
1 & 3 & 2 & 4 \\
2 & 4 & 1 & 3
\end{array}
\qquad
\begin{array}{cccc}
1 & 3 & 2 & 1 \\
2 & 1 & 3 & 2 \\
3 & 2 & 4 & 3 \\
4 & 1 & 2 & 4 \\
2 & 4 & 1 & 3
\end{array}
$$

so that a $(5 \times n; 3)$ grid can be 4-colored, by first coloring the (unshifted) $5 \times (n-4)$ grid and then adding on the necessary remaining four columns. This is valid provided $n \geq 7$.

Similarly the $(5 \times 4; 4)$ grid can be 4-colored with both starting 4-colorings so that a $(5 \times n; 4)$ grid can be 4-colored for $n \geq 7$.

$$
\begin{array}{cccc}
1 & 4 & 3 & 2 \\
2 & 1 & 4 & 3 \\
3 & 2 & 1 & 4 \\
1 & 4 & 2 & 3 \\
2 & 3 & 1 & 4
\end{array}
\qquad
\begin{array}{cccc}
1 & 3 & 2 & 1 \\
2 & 1 & 4 & 3 \\
3 & 2 & 1 & 4 \\
4 & 3 & 2 & 3 \\
2 & 4 & 1 & 4
\end{array}
$$

However, the $(5 \times 4; 5)$ grid can be 4-colored with the starting column 1, 2, 3, 4, 2, but not the starting column $1, 2, 3, 1, 2$. We can find a 4-coloring of the $(5 \times 5; 5)$ grid with starting column 1, 2, 3, 1, 2, see below. This proves that the $(5 \times n; 5)$ grid can be 4-colored provided $n \geq 8$.

$$
\begin{array}{ccccc}
1 & 3 & 2 & 1 & 4 \\
2 & 1 & 3 & 2 & 3 \\
3 & 2 & 4 & 1 & 4 \\
1 & 3 & 2 & 3 & 1 \\
2 & 4 & 1 & 4 & 2
\end{array}
\qquad
\begin{array}{cccc}
1 & 3 & 4 & 1 \\
2 & 1 & 3 & 4 \\
3 & 2 & 1 & 3 \\
4 & 3 & 2 & 4 \\
2 & 4 & 1 & 2
\end{array}
$$

It is straightforward to check that the size of the largest independent set in the $(5 \times 3; 2)$ or $(5 \times 3; 3)$ is 3; and that the size of the largest independent set in a $(5 \times 5; 3)$ or $(5 \times 5; 4)$ is less than 7, so that these grids cannot be 4-colored. We 4-color the remaining small grids not covered by our general results in an appendix at the end of the paper. $\qquad\square$

THEOREM 3.8. *Let G be a 6-regular right-diagonal $(m \times 2; i)$ grid on the torus with $1 \leq i \leq m$. Then if m is even, G can be 4-colored. If $i = 1$ and m is odd, G cannot be 4-colored. If $i > 1$ and m is odd, then G is also a $(r \times s; t)$ grid where $r, s \neq 2$, $rs = 2m$, and $1 \leq t \leq r$, and so its colorability is settled by other results.*

PROOF. As mentioned previously an unshifted $m \times 2$ grid is 4-colorable if and only if m is even. The same coloring works for a $(m \times 2; i)$ grid when m is even, namely by assigning two colors alternately to each of the vertical cycles.

Suppose $i > 1$ and m is odd. Let the vertices of $C_v(1)$ be labelled $1, 2, \ldots, m$, and those of $C_v(2)$ by $1', 2', \ldots, m'$. As explained after Altshuler's Theorem, G can be redrawn as another grid graph by following the cycle 1, $1'$, i, i', $2i - 1$, $(2i - 1)', \cdots$ until $ki - k \equiv 0 \pmod{m}$. Let $r = \gcd(i - 1, m)$. If $r = 1$, then G is represented as a $(2m \times 1; j)$ grid for some j, $1 < j \leq 2m$, and we refer to

Theorem 3.9. Otherwise since r is odd, it is at least 3, and G is represented as an $(r \times (2k); j)$ grid with r, $2k > 2$ and so is covered by Theorem 3.3, 3.6 or 3.7. $\square$

We are left with the $(m \times 1; i)$ grid cases; these are circulant graphs with vertex j adjacent (clockwise) to $j - 1$, $j + i - 2$, $j + i - 1$, $j + 1$, $j - i + 2$, and $j - i + 1$ (mod m) for $j = 1, 2, \ldots, m$, or equivalently with difference set $\{1, i - 2, i - 1\}$. The grids with $i = 1$ and 2 contain loops and so we do not consider them. The grids with $i = 3$ contain multiple edges and are circulants with vertex j adjacent to vertices $j - 2, j - 1, j + 1$, and $j + 2$ (mod m). These can be 4-colored for $m > 5$ by $abc \ldots abc$ if $m \equiv 0$ (mod 3), $abc \ldots abcabcd$ if $m \equiv 1$ (mod 3), and by $abc \ldots abcadcbd$ if $m \equiv 2$ (mod 3). In the grids with $i = 4$ every successive set of 4 vertices forms a complete graph so these graphs can be 4-colored if and only if 4 divides m. If 4 does not divide m and $m \geq 15$, it is easy to check that the grid is 5-colorable. These grids are the family of 5-chromatic, 6-regular toroidal graphs of edge-width three referred to in Section 1; the cycle $(1, 2, 3)$ is non-contractible. The remaining cases of $5 \leq i \leq m$ are more complex; due to symmetry we may assume that $i \leq \lceil m/2 \rceil$.

THEOREM 3.9. *Consider 6-regular right-diagonal $(m \times 1; i)$ grids on the torus with $5 \leq i \leq \lceil m/2 \rceil$. Then for each value of i there is a finite number of these grids that are not 4-colorable.*

For example, direct computation shows that the $(m \times 1; 5)$ grids are 4-colorable for $m \geq 26$, and are not 4-colorable only for $m = 10, 11, 13, 17, 18$ and 25. Similarly the $(m \times 1; 6)$ grids are all 4-colorable for $m \geq 18$.

PROOF. Let the vertices be labelled $0, 1, \ldots, m - 1$ so that vertex 0 is adjacent to vertices $m - 1$, $i - 2$, $i - 1$, 1, $m - i + 2$, and $m - i + 1$. Suppose that i is odd with $i = 2k + 1$ for some integer k. For $m = 4k - 1$ we color the vertices by $(ab)^k(cd)^{k-1}c$; that is, by k repetitions of ab, then $k - 1$ repetitions of cd, followed by one c. For $m = 4k$ we color with $(ab)^k(cd)^k$.

To check that this is a valid coloring, in this and subsequent cases, we list the vertex numbers in each color class and then check that no two numbers in one class differ by an element of the difference set $S = \{1, 2k - 1, 2k\}$ with all work performed (mod m). In these cases we have, respectively:

$$
\begin{array}{cccccc}
a: & 0 & 2 & \ldots & 2k - 4 & 2k - 2 \\
b: & 1 & 3 & \ldots & 2k - 3 & 2k - 1 \\
c: & 2k & 2k + 2 & \ldots & 4k - 4 & 4k - 2 \\
d: & 2k + 1 & 2k + 3 & \ldots & 4k - 3 &
\end{array}
$$

$$
\begin{array}{cccccc}
a: & 0 & 2 & \ldots & 2k - 4 & 2k - 2 \\
b: & 1 & 3 & \ldots & 2k - 3 & 2k - 1 \\
c: & 2k & 2k + 2 & \ldots & 4k - 4 & 4k - 2 \\
d: & 2k + 1 & 2k + 3 & \ldots & 4k - 3 & 4k - 1
\end{array}
$$

Note that these two colorings are "compatible," meaning that concatenating one with the other gives a valid coloring. For example, with $m = 4k - 1 + 4k = 8k - 1$, a valid coloring is given by

$$(ab)^k(cd)^{k-1}c(ab)^k(cd)^k.$$

More generally, if $m = A(4k - 1) + B(4k)$ with A and B nonnegative integers, then we can color the $(m \times 1; 2k + 1)$ grid with first the coloring for $4k - 1$ vertices,

$(ab)^k(cd)^{k-1}c$, concatenated A times, and followed with B copies of the second coloring, $(ab)^k(cd)^k$. This gives a proper coloring of all m vertices. Since $4k-1$ and $4k$ are relatively prime, all m sufficiently large can be expressed as $A(4k-1)+B(4k)$. In fact, if x and y are relatively prime integers, then each $n \geq (x-1)(y-1)$ can be expressed as $n = Ax + By$ for some nonnegative integers A and B ([**23**], see also [**17**]); thus we can 4-color all such graphs with $m \geq (4k-2)(4k-1) = (2i-4)(2i-3)$.

Next let i be even so that $i = 2k + 2$ and vertex 0 is adjacent to 1, $2k$, $2k+1$, $m-1$, $m-2k$, and $m-2k-1$ for some integer k with $i = 2k+2$; now we have the difference set $S = \{1, 2k, 2k+1\}$. We solve this in three cases, modulo 6.

First suppose that $i \equiv 0 \pmod 6$, so $2k = 6j-2$ and that $S = \{1, 6j-2, 6j-1\}$ (so $3j - 1 = k$). For every m divisible by 3, $(abc)^{(m/3)}$ is a valid coloring. For $m = 24j - 8$, we use the coloring $(abc)^{2j}(dbc)^{(2j-1)}(dac)^{(2j-1)}(dab)^{(2j-1)}d$. It is straightforward to check that these are correct colorings. Note that the colorings are compatible (can be concatenated) since the first coloring is the start of the second; even better, for any m divisible by 3, the latter coloring is compatible with $(abc)^{(m/3)}$. Since 3 and $24j - 8$ are relatively prime, these circulants on $m \geq 2(24j - 9) = 2(8k - 1) = 16k - 2 = 8i - 18$ vertices can all be 4-colored.

Next suppose that $i \equiv 2 \pmod 6$, so $6j = 2k$ and $S = \{1, 6j, 6j+1\}$. For $m = 6j + 2$ we color with $(abc)^j(adc)^jbd$. For $m = 12j + 1$ we color with

$$(abc)^j(adc)^jbd(abc)^{(j-1)}(adc)^jbd.$$

The latter coloring is compatible with the former, seen by the initial segment. Since $gcd(6j + 2, 12j + 1) = 1$, for $m \geq 12j(6j + 1) = 4k(2k + 1) = 2(i - 2)(i - 1)$ all these graphs can be 4-colored.

And finally, let $i \equiv 4 \pmod 6$, so $6j + 2 = 2k$ and $S = \{1, 6j+2, 6j+3\}$. (Every coloring of the form $(abc)^sd(abc)^td$ is valid when s, t lie in $\{2k, 2k+1\}$, but we don't need that much.) For $m = 6j + 4$ we can color with $(abc)^{(2j+1)}d$, and for $m = 12j + 5$ we color with $(abc)^{(2j+1)}d(abc)^{2j}d$. These colorings are compatible since one is the start of the other. And $gcd(6j + 4, 12j + 5) = 1$ so that for all $m \geq (6j + 3)(12j + 4) = 4k(2k + 1) = 2(i - 2)(i - 1)$ we can 4-color this form of circulant graph. $\qquad\square$

Although we would like to, we cannot at present give a bound on edge-width for the $(m \times 1; i)$ grids that implies 4-colorability. Note that in these graphs the $(i-1)$-cycle $(0, 1, 2, \ldots, i-2, 0)$ is embedded as a noncontractible cycle, as are the cycles

$$(0, i - 1, 2i - 2, \ldots, qi - q, qi - q + 1, qi - q + 2, \ldots, m - 1, 0),$$
$$(0, i - 1, 2i - 2, \ldots, qi - q, (q + 1)(i - 1), (q + 1)(i - 1) - 1,$$
$$(q + 1)(i - 1) - 2, \ldots, 1, 0)$$

where $q = \lfloor m/i - 1 \rfloor$ and numbers are read modulo m. Either $q(i-1)$ or $(q+1)(i-1)$ is within $(i - 1)/2$ of m. Thus the edge-width of the $(m \times 1; i)$ grid is at most $min\{i - 1, m/i - 1 + i - 1/2\}$.

Theorem 3.9 gives a lower bound B_i on m, such that if $m \geq B_i$, then the $(m \times 1; i)$ grid can be 4-colored. We have run computer searches for $i = 5, 6, 7, 8, 9, 10$ to establish the actual least value M_i such that every $(m \times 1; i)$ grid with $m \geq M_i$ can be 4-colored, and present these bounds below. The last column in the table is the largest edge-width in a non-4-colorable shift i grid. Additional results on coloring these and other circulants can be found in [**6**].

TABLE 1. Actual and estimated bounds on m for which an $(m \times 1; i)$ grid is 4-colorable.

i	M_i	B_i	edge-width
5	26	42	4
6	18	30	4
7	18	110	3
8	34	84	5
9	27	210	5
10	27	144	3

Coloring Appendix

We color the $(5 \times 4; 2)$ grid.

```
1 4 2 3
2 1 4 1
3 2 3 2
4 1 4 3
2 3 1 4
```

We color the $(5 \times 6; 3)$ grid.

```
1 4 2 3 1 4
2 1 4 2 3 1
3 2 3 1 2 3
4 1 2 3 1 4
2 3 1 4 2 3
```

We color the $(5 \times 3; 4)$ and the $(5 \times 6; 4)$ grids.

```
1 3 2     1 3 2 1 3 2
2 1 3     2 1 3 2 4 3
3 2 4     3 2 4 1 2 4
4 1 3     4 1 3 4 1 3
2 4 1     2 4 1 3 4 1
```

We color the $(5 \times 3; 5)$, $(5 \times 6; 5)$, and the $(5 \times 7; 5)$ grid.

```
1 4 3     1 4 2 1 4 3     1 3 2 4 3 1 4
2 1 4     2 1 4 2 1 4     2 1 3 2 4 2 3
3 2 3     3 2 3 4 2 3     3 2 4 1 3 1 4
4 1 4     4 1 2 3 1 4     1 3 2 4 2 3 1
2 3 1     2 3 1 4 2 1     2 4 1 3 1 4 2
```

References

1. M. O. Albertson and J. P. Hutchinson, *Hadwiger's conjecture and 6-chromatic toroidal graphs*, Graph Theory and Related Topics (J. A. Bondy and U.S.R. Murty, eds.) Academic Press, New York, 1979, pp. 35–40.

2. _____, *On 6-chromatic toroidal graphs*, Proc. London Math. Soc. **41** (1980), 533–556.

3. M. O. Albertson and W. Stromquist, *Locally planar toroidal graphs are 5-colorable*, Proc. Amer. Math. Soc. **84** (1982), 449–456.

4. A. Altshuler, *Construction and enumeration of regular maps on the torus*, Discrete Math. **4** (1973), 201–217.

5. K. Appel and W. Haken, *Every planar map is four colorable*, Bull. Amer. Math. Soc. **82** (1976), 711–712.

6. K. L. Collins, D. Fisher, and J. P. Hutchinson, *On 3- and 4-coloring some circulant graphs*, in preparation.

7. S. Fisk, *The nonexistence of colorings*, J. Combin. Theory Ser. B, **24** (1978), 247–248.

8. M. R. Garey and D. S. Johnson, *Computers and intractability*, W. H. Freeman and Co., San Francisco, CA, 1979.

9. N. Hartsfield and G. Ringel, *Minimal quadrangulations of orientable surfaces*, J. Combin. Theory Ser. B **46** (1989), 84–95.

10. P. J. Heawood, *Map-colour theorem*, Quart. J. Pure Appl. Math. **24** (1890), 332–338.

11. _____, *On the four-colour map theorem*, Quart. J. Pure Appl. Math. **29** (1898), 270–285.

12. J. P. Hutchinson, *On coloring maps made from Eulerian graphs*, Proc. 5th British Combinatorial Conf. (Aberdeen, 1975), Congr. Numer., no. 15, Utilitas Math., Winnipeg, Man., 1976, pp. 343–354.

13. _____, *Three-coloring graphs embedded on surfaces with all faces even-sided*, J. Combin. Theory Ser. B **65** (1995), 139–155.

14. J. P. Hutchinson and B. Richter, personal communication.

15. T. R. Jensen and B. Toft, *Graph coloring problems*, John Wiley & Sons, New York, 1995.

16. A. B. Kempe, *On the geographical problem of the four colours*, Amer. J. Math. **2** (1879), 193–200.

17. A. Nijenhuis and H. S. Wilf, *Representations of integers by linear forms in nonnegative integers*, J. Number Theory **4** (1972), 98–106.

18. G. Ringel, *Map color theorem*, Springer-Verlag, Berlin, 1974.

19. G. Ringel and J. W. T. Youngs, *Solution of the Heawood map-coloring problem*, Proc. Nat. Acad. Sci. U.S.A. **60** (1968), 438–445.

20. N. Robertson, D. Sanders, P. Seymour, and R. Thomas, *The four colour theorem*, J. Combin. Theory Ser. B **70** (1997), 2–44.

21. R. Steinberg, *The state of the three color problem*, Quo Vadis, Graph Theory? (J. Gimbel, J. W. Kennedy, and L. V. Quintas, eds.), Ann. Discrete Math., vol 55, North-Holland, Amsterdam, 1993, pp. 211–248.

22. _____, *An update on the state of the three color problem*, Graph Theory Notes of New York XXV (J. W. Kennedy and L. V. Quintas, eds.), New York Academy of Sciences, New York, 1993.

23. J. Sylvester, *Mathematical questions, with their solutions*, Educational Times, vol. 41, 1884.

24. C. Thomassen, *Embeddings of graphs with no short noncontractible cycles*, J. Combin. Theory Ser. B **48** (1990), 155–177.

25. _____, *Tilings of the torus and the Klein bottle and vertex-transitive graphs on a fixed surface*, Trans. Amer. Math. Soc. **323** (1991), 605–635.

26. _____, *Five-colorings maps on surfaces*, J. Combin. Theory Ser. B **59** (1993), 89–105.

27. _____, *Trees in triangulations*, J. Combin. Theory Ser. B **60** (1994), 56–62.

28. _____, *Five-coloring graphs on the torus*, J. Combin. Theory Ser. B **62** (1994), 11–33.

29. D. B. West, *Introduction to graph theory*, Prentice Hall, 1996.

30. D. A. Youngs, *4-chromatic projective graphs*, J. Graph Theory **21** (1996), 219–227.

DEPT. OF MATHEMATICS, WESLEYAN UNIVERSITY, MIDDLETOWN, CT 06459, USA
E-mail address: kcollins@wesleyan.edu

DEPT. OF MATHEMATICS AND COMPUTER SCIENCE, MACALESTER COLLEGE, ST. PAUL, MN 55105, USA
E-mail address: hutchinson@macalester.edu

Centre de Recherches Mathématiques
CRM Proceedings and Lecture Notes
Volume **23**, 1999

On the Complexity of a Restricted List-Coloring Problem

Moshe Dror, Gerd Finke, Sylvain Gravier, and Wieslaw Kubiak

ABSTRACT. We investigate a restricted list-coloring problem: for a given graph $G = (V, E)$ and a given list L on V, we want to find an L-coloring of G such that the size of each class of colors is equal to a given integer. This restricted list-coloring problem was proposed by de Werra. We prove that this problem is $\mathcal{NP}$-Complete even if the graph is a path with at most two colors on each vertex list. We then give a polynomial algorithm to solve the special case of this problem where the total number of colors occurring in all lists is fixed.

1. Introduction

We consider finite, undirected graphs without loops. In a graph $G = (V, E)$ a k-*coloring* is a mapping $c \colon V \to \{1, 2, \ldots, k\}$ such that $c(u) \neq c(v)$ for every edge uv. Each color class is a stable set, hence a k-coloring can be seen as a partition of V into stable sets $S_1, \ldots, S_k$. The graph G is called k-colorable if it admits a k-coloring. Vizing [**6**], as well as Erdős *et al.* [**2**] introduced the following variant of the coloring problem. Suppose that each vertex v is assigned a list $L(v)$ of possible colors, and let $L(V)$ be the set of all colors. We then want to find a vertex coloring c such that $c(v) \in L(v)$ for all $v \in V$. If such a c exists, we say that the graph G is L-colorable, or that c is an L-coloring of G.

The problem to decide if graph G is L-colorable for given L is NP-complete, even when G is a complete bipartite graph [**3**].

We investigate the following restricted list-coloring problem (G, L, p), where $G = (V, E)$ is a graph, L is a list of colors on V, and p is a mapping which assigns a positive integer $p(\alpha)$ to each color $\alpha \in L(V)$. We call p a color-mapping. We then want to find an L-coloring c of G such that the color α occurs $p(\alpha)$ times, i.e. $\left| c^{-1}(\alpha) \right| = p(\alpha)$.

De Werra [**7**] proved that the problem (G, L, p) is NP-complete if G is the line-graph of a bipartite graph and for all vertex v in G, we have $\left| L(v) \right| \geq 2$. He also gave a polynomial-time algorithm to solve (G, L, p) if G is a union of disjoint cliques. In the same paper, de Werra conjectured the following:

1991 *Mathematics Subject Classification.* 05C20, 05C38.

The research of the third was supported in part by the Eőtvős University Múzeum kőrút 6-8. Budapest H-1088, Hungary.

Reprinted from Discrete Math. **195** (1999), 103–109, ⓒ1999 with permission from Elsevier Science.

CONJECTURE 1. [7] The problem (P_n, L, p) is NP-complete, where P_n is a path of length n.

In Section 3, we prove that Conjecture 1 holds. In Section 4, we give a polynomial-time algorithm for the following more restricted problem (P_n, L, p, k) where k, not being a part of the problem's input, denotes the total number of different colors available:

INSTANCE. P_n is a path of length n, L is a list of colors on V with $|L(V)| \leq k$, and p is a color-mapping.

QUESTION. Is there an L-coloring c of G such that $|c^{-1}(\alpha)| = p(\alpha)$ for all $\alpha \in L(V)$?

Xu [8] proposes a polynomial time algorithm for problem $(P_n, L, p, 3)$. In Section 4, we give a polynomial algorithm based on a dynamic programming approach which solves problem (P_n, L, p, k), for any $k \geq 3$.

2. Preliminary Results

We begin this section by showing that from the complexity point of view there is not much difference between graph G being a single path and G being a union of disjoint paths. This simple observation will make our discussion in the subsequent sections easier.

Let $G = \bigcup P_{n_i}$ be a graph which is a union of r disjoint paths P_{n_i} of length n_i. Define index$(i) = \sum_{j=1}^{i} n_j$. Let path $P_{n_i} = x_{1+\text{index}(i-1)}, \ldots, x_{n_i+\text{index}(i-1)}$. Also, let L be an assignment of lists of colors to the vertices in V, and p be a mapping on $L(V)$ to the set of positive integers. We construct a path P_n of length $r - 1 + \sum_{i=1}^{r} n_i$ by adding $r - 1$ dummy vertices $y_1, \ldots, y_{r-1}$ where y_i links path P_{n_i} to $P_{n_{i+1}}$ by edges $x_{\text{index}(i)}y_i$ and $y_i x_{1+\text{index}(i)}$. Furthermore, we set $L'(y_i) = \{\beta\}$, $p'(\beta) = r - 1$ where β is a color which does not appear in $L(V)$, $L'(v) = L(v)$ for all $v \in V$, and $p'(\alpha) = p(\alpha)$ for all $\alpha \in L(V)$.

LEMMA 1. *The problems $(\bigcup P_{n_i}, L, p)$ and (P_n, L', p') are equivalent.*

PROOF. Since $\beta \notin L(V)$, it is obvious that there exists a solution to (P_n, L', p') if and only if there exists a solution to $(\bigcup P_{n_i}, L, p)$. Given a solution to (P_n, L', p') one can easily obtain a solution to $(\bigcup P_{n_i}, L, p)$ by simply dropping color β. Conversely, given a solution to $(\bigcup P_{n_i}, L, p)$ one can easily obtain a solution to (P_n, L', p') by simply adding color β. □

In our proof of Conjecture 1 we shall use the following NP-complete problem POS-EXACT-3SAT [1]:

INSTANCE. Set U of variables, collection $\mathcal{C}$ of clauses over U such that

1. each clause $c \in \mathcal{C}$ has size 3,
2. each variable of U appears in exactly three distinct clauses,
3. no $c \in \mathcal{C}$ contains a negated literal.

QUESTION. Is there a truth assignment for U such that each clause in $\mathcal{C}$ has exactly one true literal?

THEOREM 1. [1] *The problem* POS-EXACT-3SAT *is NP-complete.*

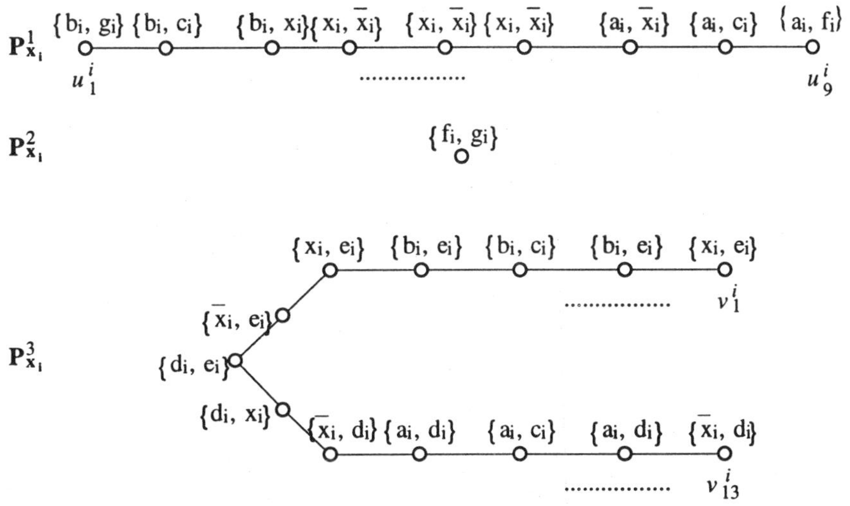

FIGURE 1. Paths $P_{x_i}^1$, $P_{x_i}^2$ and $P_{x_i}^3$.

3. $\mathcal{NP}$-Completeness of $(\bigcup P_{n_i}, L, p)$

THEOREM 2. *There exists a polynomial reduction from* POS-EXACT-3SAT *to* $(\bigcup P_{n_i}, L, p)$.

PROOF. We give a linear reduction from POS-EXACT-3SAT to $(\bigcup P_{n_i}, L, p)$ problem.

For each variable x_i we define "gadget" GX_i (see Fig. 1), which is a union of three disjoint paths $P_{x_i}^j$ for $j = 1, 2$ and 3, where:

$P_{x_i}^1$ is a path $u_1^i, \ldots, u_9^i$ of 9 vertices, with their lists $L(u_1^i) = \{b_i, g_i\}$, $L(u_2^i) = \{b_i, c_i\}$, $L(u_3^i) = \{b_i, x_i\}$, $L(u_4^i) = L(u_5^i) = L(u_6^i) = \{x_i, \overline{x}_i\}$, $L(u_7^i) = \{a_i, \overline{x}_i\}$, $L(u_8^i) = \{a_i, c_i\}$ and $L(u_9^i) = \{a_i, f_i\}$.

$P_{x_i}^2$ is a vertex w^i with its list $\{f_i, g_i\}$.

$P_{x_i}^3$ is a path $v_1^i, \ldots, v_{13}^i$ of 13 vertices, with their lists $L(v_1^i) = L(v_5^i) = \{x_i, e_i\}$, $L(v_2^i) = L(v_4^i) = \{b_i, e_i\}$, $L(v_3^i) = \{b_i, c_i\}$, $L(v_6^i) = \{\overline{x}_i, e_i\}$, $L(v_7^i) = \{d_i, e_i\}$, $L(v_8^i) = \{d_i, x_i\}$, $L(v_9^i) = L(v_{13}^i) = \{\overline{x}_i, d_i\}$, $L(v_{10}^i) = L(v_{12}^i) = \{a_i, d_i\}$ and $L(v_{11}^i) = \{a_i, c_i\}$.

For each clause $C_w = x_i + x_j + x_k \in \mathcal{C}$ we define three paths $P_{C_w}^1$, $P_{C_w}^2$, and $P_{C_w}^3$ where $P_{C_w}^1$ is a single vertex c_w^1 with its list $\{x_i, x_j, x_k\}$ of colors, $P_{C_w}^2$ and $P_{C_w}^3$ are single vertices c_w^2 and c_w^3, respectively, with the same list $\{\overline{x}_i, \overline{x}_j, \overline{x}_k\}$ of colors for each.

We have the following, crucial for our reduction, property of the "gadgets".

LEMMA 2. *For any L-coloring c of GX_i with $\left|c^{-1}(a_i)\right| = \left|c^{-1}(b_i)\right| = \left|c^{-1}(d_i)\right| = \left|c^{-1}(e_i)\right| = 3$, $\left|c^{-1}(f_i)\right| = \left|c^{-1}(g_i)\right| = 1$ and $\left|c^{-1}(c_i)\right| = 2$, we have either $\left|c^{-1}(x_i)\right| = 2$ and $\left|c^{-1}(\overline{x}_i)\right| = 5$ or $\left|c^{-1}(\overline{x}_i)\right| = 2$ and $\left|c^{-1}(x_i)\right| = 5$.*

PROOF. Consider an L-coloring c of GX_i satisfying the hypothesis of Lemma 2. Assume that $c(u_5^i) = x_i$, and so $c(u_4^i) = c(u_6^i) = \overline{x}_i$, $c(u_7^i) = a_i$ and $c(u_8^i) = c_i$.

First, we show by contradiction that $c(v_{11}^i) = a_i$. Suppose that $c(v_{11}^i) = c_i$. Thus, since $\left|c^{-1}(c_i)\right| = 2$, we have $c(v_3^i) = c(u_2^i) = b_i$ which implies that $c(u_1^i)$, $c(u_3^i)$, $c(v_2^i)$ and $c(v_4^i)$ must differ from b_i. Therefore we end up with being short of color b_i, i.e. $\left|c^{-1}(b_i)\right| = 2$, and we get a contradiction.

Thus $c(v_{11}^i) = a_i$, and consequently, $c(v_{10}^i) = c(v_{12}^i) = d_i$, $c(v_9^i) = c(v_{13}^i) = \overline{x}_i$. Since $\left|c^{-1}(a_i)\right| = 3$, we need one more vertex colored a_i. The only not colored yet vertex y in GX_i with color a_i on its list $L(y)$ is u_9^i, thus, we must have $c(u_9^i) = a_i$. Then, since $\left|c^{-1}(f_i)\right| = \left|c^{-1}(g_i)\right| = 1$, we have $c(u_1^i) = g_i$ and $c(w^i) = f_i$.

Consider color b_i. We observe that the only vertices with color b_i on their lists in GX_i still available for coloring are u_2^i, u_3^i, v_2^i, v_3^i and v_4^i. We have that at most one of two vertices u_2^i and u_3^i can be colored b_i, and that at most two of three vertices v_2^i, v_3^i and v_4^i can be colored b_i. Since $\left|c^{-1}(b_i)\right| = 3$, then the only possibility is to have $c(v_2^i) = c(v_4^i) = b_i$, which implies that $c(v_3^i) = c_i$. Since $\left|c^{-1}(c_i)\right| = 2$, we obtain that $c(u_2^i) = b_i$, and consequently $c(u_3^i) = x_i$.

Now, consider color e_i. We observe that the only vertices with color e_i on their lists in GX_i still available for coloring are v_1^i, v_5^i, v_6^i and v_7^i. Since $\left|c^{-1}(e_i)\right| = 3$, then we have $c(v_1^i) = c(v_5^i) = c(v_7^i) = e_i$, and consequently $c(v_6^i) = \overline{x}_i$. Since $\left|c^{-1}(d_i)\right| = 3$, we have $c(v_8^i) = d_i$.

Finally, we reached the point where c is completely determined. It is an L-coloring of GX_i which satisfy the hypothesis of Lemma 2 with $c^{-1}(x_i) = \{u_3^i, u_5^i\}$ and $c^{-1}(\overline{x}_i) = \{u_4^i, u_6^i, v_6^i, v_9^i, v_{13}^i\}$.

To finish the proof, we notice that the case with $c(u_5^i) = \overline{x}_i$ holds by the symmetry of x_i and $\overline{x}_i$ in "gadget" GX_i.   □

Lemma 2 suggests the following mapping p to complement our reduction. For each variable $x_i \in U$, we set $p(x_i) = p(\overline{x}_i) = 5$, $p(a_i) = p(b_i) = p(d_i) = p(e_i) = 3$, $p(f_i) = p(g_i) = 1$ and $p(c_i) = 2$.

If x_i is true, then we choose the coloring c of GX_i with color x_i appearing twice and color $\overline{x}_i$ five times, and $p(a_i) = p(b_i) = p(d_i) = p(e_i) = 3$, $p(f_i) = p(g_i) = 1$ and $p(c_i) = 2$. By definition of POS-EXACT-3SAT, there exist three clauses C_w where the variable x_i appears and for each of these clauses x_i is the only true literal under the assignment, so we can color vertices c_w^1 with color x_i. Such a coloring c satisfies $\left|c^{-1}(x_i)\right| = p(x_i)$.

Now, if x_i is false, then we choose the coloring c of GX_i with color $\overline{x}_i$ appearing twice and color x_i five times, and $p(a_i) = p(b_i) = p(d_i) = p(e_i) = 3$, $p(f_i) = p(g_i) = 1$ and $p(c_i) = 2$. By definition of POS-EXACT-3SAT, there exist three clauses C_w where the variable x_i appears and each of these clauses has exactly two false literals, so we can color either vertex c_w^2 or vertex c_w^3 with color $\overline{x}_i$. Such a coloring c satisfies $\left|c^{-1}(\overline{x}_i)\right| = p(\overline{x}_i)$.

Since X is a truth assignment to the variable in U such that each clause has exactly one true literal, then exactly $|U|/3$ variables must be true and exactly $2|U|/3$ must be false in X. Thus, the coloring c colors all $3|U|$ single vertex paths $P_{C_w}^1$, $P_{C_w}^2$, and $P_{C_w}^3$, $w = 1, \dots, |U|$, and, therefore, it is a complete and feasible coloring of the graph

$$\{\{P_{C_w}^1, P_{C_w}^2, P_{C_w}^3\}_{C_w \in \mathcal{C}} \cup \{P_{x_i}^1, P_{x_i}^2, P_{x_i}^3\}_{x_i \in U}\}.$$

Now assume that there exists an L-coloring c of the graph

$$\left\{\{P_{C_w}^1, P_{C_w}^2, P_{C_w}^3\}_{C_w \in \mathcal{C}} \cup \{P_{x_i}^1, P_{x_i}^2, P_{x_i}^3\}_{x_i \in U}\right\}.$$

Consider the following assignment defined by c: if exactly two vertices of GX_i are colored x_i then set $x_i := true$, otherwise, set $\bar{x}_i := true$. By Lemma 2, this assignment is a correct truth assignment to the variables in U.

Assume that x_i is true. Then color x_i appears exactly twice in GX_i. Thus, since a variable x_i occurs in exactly three clauses C_w and $p(x_i) = 5$, we must have $c(c_w^1) = x_i$ for each of these clauses. Thus each clause has exactly one true literal under the assignment defined by c. $\qquad\square$

REMARK. If for all variables x_i in U we add: eight vertices with their lists $\{a_i\}$, $\{a_i\}$, $\{b_i\}$, $\{b_i\}$, $\{d_i\}$, $\{d_i\}$ and $\{e_i\}$, $\{e_i\}$; eight vertices with their lists $\{f_i\}, \{f_i\}, \{f_i\}, \{f_i\}$ and $\{g_i\}$, $\{g_i\}$, $\{g_i\}$, $\{g_i\}$; and three vertices with the same list $\{c_i\}$, then our reduction will hold even for a constant function $p = 5$.

The following corollary is a direct consequence of Theorems 1, 2 and Lemma 1.

COROLLARY 1. *Conjecture 1 holds.*

We can even strengthen this result by showing the following theorem.

THEOREM 3. *Problem* $(\bigcup P_{n_i}, L, p)$ *remains* NP-*complete even if* $|L(v)| \leq 2$ *for every* $v \in V$.

PROOF. The proof takes the reduction from Theorem 2 as its point of departure. In the original instance I of $(\bigcup P_{n_i}, L, p)$ built by this reduction, we have that $|L(v)| \leq 3$ for every $v \in V$, and any vertex v with $|L(v)| = 3$ makes a *single vertex* path. Now, let $v_1, \ldots, v_m$ be all vertices with three-color lists assigned to them by the reduction. Let $L(v_i) = \{A_i, B_i, C_i\}$, for $i = 1, \ldots, m$. Replace path v_i by three new single vertex paths v_i^1, v_i^2, and v_i^3 with *two*-color lists $L(v_i^1) = \{A_i, T_i\}$, $L(v_i^2) = \{B_i, T_i\}$, and $L(v_i^3) = \{C_i, T_i\}$, where color T_i is not in $L(V)$. Furthermore, set $p(T_i) = 2$. Do the above for $i = 1, \ldots, m$, and set $L'(V) := L(V) \bigcup \{T_1, \ldots, T_m\}$ to make a new instance I'. No other changes are made to the original I.

Notice that any L'-coloring for I' colors exactly two out of three vertices v_i^1, v_i^2, and v_i^3 with color T_i. Thus exactly one of them is colored with a color from $L(V)$, $i = 1, \ldots, m$. This color can then be chosen to color vertex v_i of the original instance I. On the other hand, any L-coloring for I chooses exactly one color from list $\{A_i, B_i, C_i\}$ to color v_i. This color uniquely defines a vertex v_i^j, $j = 1, 2, 3$, of I' which should be colored with it. The other two vertices v_i^j, $j = 1, 2, 3$ are colored with T_j. Thus by Theorems 1 and 2, and Lemma 1 the theorem holds. $\qquad\square$

4. A Polynomial Algorithm for (P_n, L, p, k)

In this section we show a polynomial-time algorithm for problem $(\bigcup P_{n_i}, L, p, k)$. This algorithm also solves problem (P_n, L, p, k) as a special single path case of the former.

Let k be a positive integer. Let $G = \bigcup P_{n_i}$ be a graph union of disjoint r paths $P_{n_i} = x_{1+\text{index}(i-1)}, \ldots, x_{n_i+\text{index}(i-1)}$, L be a list of colors on vertex set V of G such that $L(V) \leq k$, and p be a mapping on $L(V)$ to the set of positive integers. Let $n = \sum_{i=1}^r n_i$.

We construct an auxiliary digraph $D = (A, H)$ where the vertices are vectors of the form

$$(1) \qquad (i, c, \lambda_1, \ldots, \lambda_k) \text{ with } \sum_{j=1}^{k} \lambda_j = i,$$

where $c \in L(x_i)$, $0 \le \lambda_j \le p(j)$ for every $j = 1, \ldots, k$ and $\lambda_c \ge 1$.

The value of λ_r in the vector $(i, c, \lambda_1, \ldots, \lambda_k)$ is going to be the number of occurrences of the colour r among the vertices $x_1, \ldots, x_i$.

We have an arc XY between $X = (i, c, \lambda_1, \ldots, \lambda_k)$ and $Y = (j, c', \lambda'_1, \ldots, \lambda'_k)$ if and only if all of the following conditions hold

1. $j = i + 1$,
2. if $x_i x_j$ is an edge of G then $c' \in L(x_j) - \{c\}$, otherwise $c' \in L(x_j)$,
3. for every $r \ne c'$ we have $\lambda_r = \lambda'_r$, and $\lambda'_{c'} = \lambda_{c'} + 1$.

Furthermore, we add *initial* vector $X0 = (0, 0, \ldots, 0)$ and *goal* vector $XN = (n + 1, 0, p(1), \ldots, p(k))$. The initial vector $X0$ is linked by arcs to all vectors of the form $(1, c, \lambda_1 \ldots, \lambda_k)$, where $\lambda_r = 0$ for $r \ne c$ and $\lambda_c = 1$. The goal vector XN is entered by arcs from all vectors of the form $(n, c, p(1) \ldots, p(k))$.

We claim that

$$(2) \qquad \begin{array}{l} \text{There exists an } L\text{-coloring } c \text{ of } G \text{ satisfying } \left|c^{-1}(\alpha)\right| = p(\alpha) \text{ for all} \\ \alpha \in L(V) \text{ if and only if there exists a path between } X0 \text{ and } XN. \end{array}$$

Indeed, first assume that we have an L-coloring c of G satisfying $\left|c^{-1}(\alpha)\right| = p(\alpha)$ for all $\alpha \in L(V)$. For all $i = 1, \ldots, n$, let $Xi = (i, c(x_i), \lambda_1 \ldots, \lambda_k)$ with $\lambda_r = \left|c^{-1}(r) \cap \{x_1, \ldots, x_i\}\right|$ for $r = 1, \ldots, k$. Then, the sequence $X0, X1, \ldots, Xn, XN$ is a path between $X0$ and XN.

Now, assume that we have a path $X0, X1, \ldots, Xt, XN$. By definition of H, we have $t = n$. Let c be L-coloring of G defined by $c(x_i) = \alpha$ if and only if the second coordinate of the vector Xi is α. By definition of XN, it is easy to see that c satisfies $\left|c^{-1}(\alpha)\right| = p(\alpha)$ for all $\alpha \in L(V)$.

Finding a path between two vertices in a digraph can be done in time linear in the number of arcs (for example, depth-first search and breadth-first search achieve this bound if adjacency lists are used to represent the digraph.) By definition of A and (1), we obtain that $|A| \le nkn^{k-1} + 2$. Since the maximum out-degree in H is at most k, we have $|H| \le k(nkn^{k-1} + 2)$. Thus our algorithm runs in time $\mathcal{O}(n^k)$.

5. Conclusion

All our results can be extended to the problem of finding an L-coloring of a union of disjoint paths where the size of the color classes are just *bounded*. Indeed, it is sufficient to add some vertices with all colors in their lists such that the number of vertices in this new graph is exactly the sum of the bounds.

Polynomial-time algorithms for (G, L, p, k), where G is a tree, were recently given independently by Gravier, Kobler and Kubiak [4] and by Jensen [5].

References

1. J. Blazewicz, J. Barcello, W. Kubiak, and H. Röck, *Scheduling tasks on two processors with deadlines and additional resources*, European J. Oper. Res. **26** (1986), 364–370.
2. P. Erdös, A. L. Rubin, and H. Taylor, *Choosability in graphs*, Proc. West Coast Conference on Combinatorics, Graph Theory, and Computing (Arcata, 1979), Congress. Numer. XXVI, Utilitas Math., Winnipeg, Manitoba, 1980, pp. 125–157.

3. S. Gravier, *Coloration et produits de graphes*, Ph.D., Univ. Joseph Fourier, Grenoble, France, 1996.
4. S. Gravier, D. Kobler, and W. Kubiak, *Complexity of list coloring problems with a fixed total number of colors* (submitted).
5. K. Jensen, private communication, 1997.
6. V. G. Vizing, *Vertex colorings with given colors*, Metody Diskret. Analiz. **29** (1976), 3–10 (Russian).
7. D. de Werra, *Some combinatorial models for course scheduling*, EPFL, report ORWP 95/09, Lausanne.
8. K. Xu, *A special case of edge-chromatic scheduling problem*, EPFL, report ORWP 96/03, Lausanne.

UNIVERSITY OF ARIZONA, MIS DEPARTMENT, COLLEGE OF BUSINESS AND PUBLIC ADMINISTRATION, TUCSON, ARIZONA 85721, USA.
   E-mail address: `mdror@bpa.arizona.edu`

UNIVERSITÉ JOSEPH FOURIER, LABORATOIRE LEIBNIZ, 46, AVENUE FÉLIX VIALLET 38031 GRENOBLE CEDEX 1, FRANCE.
   E-mail address: `gerd.finke@imag.fr`

UNIVERSITÉ JOSEPH FOURIER, LABORATOIRE LEIBNIZ, 46, AVENUE FÉLIX VIALLET 38031 GRENOBLE CEDEX 1, FRANCE.
   E-mail address: `sylvain.gravier@imag.fr`

ÉCOLE NATIONALE SUPÉRIEURE DE GÉNIE INDUSTRIEL, LABORATOIRE GILCO, AND UNIVERSITÉ JOSEPH FOURIER, LABORATOIRE LEIBNIZ, 46, AVENUE FÉLIX VIALLET 38031 GRENOBLE CEDEX 1, FRANCE.
   Current address: Faculty of Business Administration, Memorial University of Newfoundland, St. John's, Newfoundland, Canada.
   E-mail address: `wieslaw.kubiak@imag.fr`

Centre de Recherches Mathématiques
CRM Proceedings and Lecture Notes
Volume **23**, 1999

Totally Critical Graphs and the Conformability Conjecture

G. M. Hamilton, A. J. W. Hilton, and H. R. F. Hind

ABSTRACT. The Conformability Conjecture of Chetwynd and Hilton, if true, would determine the total chromatic number of all graphs G with $\Delta(G) > |V(G)|/2$, a class of graphs to which nearly all graphs belong. The conjecture is first modified to take the exceptional graphs of Chen and Fu into account, and this modification is justified in some detail. For graphs of even order that are totally critical (that is, critical with respect to the total chromatic number), it is shown that the Conformability Conjecture takes a spectacularly simple form. If this simple form of the Conformability Conjecture is correct, then it would be possible to test in polynomial time whether an even order graph G with $\Delta(G) > |V(G)|/2$ is totally critical; for even order graphs G satisfying $|V(G)|/2 < \Delta(G) \leq 3|V(G)|/4 - 1$ it would also follow that if G were totally critical, then G would be regular.

The Conformability Conjecture is established for odd order graphs G satisfying $\Delta(G) \geq \{\sqrt{7}|V(G)|+\mathrm{def}(G)+1\}/3$ and $\mathrm{def}(G) \leq |V(G)|-\Delta(G)-1$, where $\mathrm{def}(G) = \Delta(G)|V(G)| - 2|E(G)|$.

The Biconformability Conjecture is also discussed, as are supercritical graphs (that is, graphs which are totally critical but have no totally critical subgraphs of maximum degree one less). A complete list of all totally critical graphs of order at most ten is given, and some open problems are stated.

1. Introduction

Of the three classical graph colouring parameters, the chromatic number $\chi(G)$, the chromatic index $\chi'(G)$ and the total chromatic number $\chi''(G)$, the total chromatic number remains the most mysterious. It is now over thirty years since Behzad [3] and Vizing [37] independently posed the total chromatic number conjecture (TCC) that, for a simple graph G,

$$\Delta(G) + 1 \leq \chi''(G) \leq \Delta(G) + 2,$$

where $\Delta(G)$ is the maximum degree of G.

1991 *Mathematics Subject Classification.* Primary: 05C15.

We would like to thank Odile Marcotte for a large number of perceptive and helpful comments in the final preparation of this paper. The second and third authors wish to acknowledge the support received from the United Kingdom Science and Engineering Research Council (Grant No. GR/F42034).

Unfortunately, G. M. Hamilton has passed away before the publication of this article.

This is the final version of the paper.

Although this conjecture remains open, it was proved by Hilton and Hind [27] that it is true for graphs G satisfying $\Delta(G) \geq (3/4)|V(G)|$, and by Kostochka [30] that it is true if $\Delta(G) \leq 5$. Moreover, McDiarmid and Reed [32] showed that the proportion of graphs of order n that do not satisfy the TCC must be very small. Three good upper bounds for $\chi''(G)$ are known; two of them are due to Hind [28, 29], namely

$$\chi''(G) \leq \chi'(G) + 2\lceil \sqrt{\chi(G)} \rceil$$

and

$$\chi''(G) \leq \chi'(G) + 2\left\lceil \frac{|V(G)|}{\Delta(G)} \right\rceil + 1,$$

and a third is due independently to McDiarmid and Reed [32] and (in a slightly stronger version) Chetwynd and Häggkvist [11]. The stronger version is that if $|V(G)| \leq t!$ then

$$\chi''(G) \leq \chi'(G) + t.$$

Very recently, Molloy and Reed [33] showed that there is a constant c such that $\chi''(G) \leq \Delta(G) + c$ for all graphs G.

In this paper we concern ourselves with another aspect of the total chromatic number conundrum. Call a simple graph G *Type* 1 if

$$\chi''(G) = \Delta(G) + 1$$

and *Type* 2 if

$$\chi''(G) \geq \Delta(G) + 2.$$

We are interested in classifying graphs as to whether they are Type 1 or Type 2. Fairly recently Sánchez-Arroyo [35] showed that the problem of determining the total chromatic number of a graph is NP-hard. In view of this it seems unlikely that a good classification exists which is valid for all graphs.

The classification that we consider is enunciated as the Conformability Conjecture in Section 2, and it only applies to graphs satisfying $\Delta(G) > |V(G)|/2$. By developing the material in Chapter 9 of [7] somewhat, it follows that almost all unlabelled graphs of fixed order, where each graph is considered equiprobable, satisfy the inequality $\Delta(G) > |V(G)|/2$ (this is also shown in [19]). Thus the Conformability Conjecture, if true, would determine the total chromatic number of all graphs in a natural class of graphs, a class to which nearly all graphs belong. While McDiarmid and Reed showed in [32] that almost all graphs are Type 1, their result does not give a method for deciding whether a given graph is Type 1 or Type 2.

The Conformability Conjecture has as its motivation the belief that, provided the vertices of a graph G with $\Delta(G) \geq |V(G)|/2$ can be coloured in a way that is not obviously incompatible with a Type 1 total colouring of G, then in fact a Type 1 total colouring of G does exist. Such a vertex-colouring we call a conformable vertex-colouring. It is not clear whether there is a polynomial algorithm to decide whether a graph has a conformable vertex colouring. Thus it is not clear whether the Conformability Conjecture would provide some kind of quasiparadoxical contrast to Sánchez-Arroyo's result.

A graph G will be called *totally critical* (or simply *critical*) if G is connected Type 2, and $\chi''(G\backslash e) < \chi''(G)$ for each edge $e \in E(G)$. In Section 3 we show that if the Conformability Conjecture is true for even order graphs, then there is an astonishingly simple criterion for determining whether or not G is critical. This would imply the existence of a polynomial time algorithm to decide whether an even order graph G is critical. Unfortunately we do not know at this point whether there is any such simple criterion to decide whether an odd order graph is conformable. We find this contrast between even order and odd order graphs very surprising. We also show in Section 3 that if the Conformability Conjecture is true then, curiously, all even order critical graphs G satisfying $|V(G)|/2 < \Delta(G) \leq 3|V(G)|/4 - 1$ are regular.

Although we have been able to render the Conformability Conjecture into a remarkably simple form when $|V(G)|$ is even, we do not have all that many firm results to support the Conjecture in this case. When $|V(G)|$ is odd the situation is quite different. In this case we cannot find any simplified form for the Conformability Conjecture, but we have been able to find substantial evidence to support the Conjecture. In particular, we have verified it for odd order graphs G satisfying

$$\Delta(G) \geq \tfrac{1}{3}\{\sqrt{7}|V(G)| + \mathrm{def}(G) + 1\}$$

and

$$\mathrm{def}(G) \leq |V(G)| - \Delta(G) - 1,$$

where $\mathrm{def}(G)$, the *deficiency* of G, is defined by

$$\mathrm{def}(G) = \sum_{v \in V(G)} \big(\Delta(G) - d_G(v)\big).$$

We show this in Section 4.

In Section 5 we discuss the total chromatic number of bipartite graphs. It is easy to see that bipartite graphs satisfy the TCC, and the interest here lies in the question of whether it is possible to classify these according to type. Sánchez-Arroyo's result [35] actually shows that the problem of determining whether a bipartite graph is Type 1 or Type 2 is NP-hard. If n is even the graph $K_{n,n}$ is conformable and yet $K_{n,n}$ is Type 2, so conformability is not the important criterion here. However there is a constraint that the vertex-colouring of a bipartite graph must satisfy if it is to be Type 1. We call it biconformability. We formulate the Biconformability Conjecture. It follows from results in [19] that the Biconformability Conjecture, if true, would classify almost all bipartite graphs. However, it is not clear whether there is a polynomial algorithm to determine if a bipartite graph is biconformable. The correct definition of biconformability has only recently been arrived at (if in fact we have really found it) and so there is less progress to report on the Biconformability Conjecture. Nonetheless the results obtained thus far are not at all trivial.

In Section 6 we draw attention to a surprising contrast between the behaviour of graphs which are critical with respect to the edge-chromatic number (edge-critical graphs) and totally critical graphs. It is well known (see [21]) that each edge-critical graph G contains an edge-critical subgraph of maximum degree one less. Much to our surprise, we discovered that the same is not true for totally critical graphs. We call graphs that are totally critical but contain no totally critical subgraphs of maximum degree one less *supercritical*. We have firm examples of supercritical graphs, and, if the Conformability Conjecture is correct, then it follows that all even order critical graphs satisfying

$$\tfrac{1}{2}|V(G)| + 1 \leq \Delta(G) \leq \tfrac{3}{4}|V(G)| - 1$$

are supercritical.

In Section 7 we pose some open problems. In Section 8 we give a list of all totally critical graphs of order at most 10; we also show that two infinite classes of graphs, suggested by the list, are critical.

Finally, let us say something about our terminology and notation. The graphs we consider in this paper are all finite simple graphs, except where it is explicitly stated that we are considering a multigraph. Given a graph G, a *colouring* is a function $\varphi\colon X \to C$, where C is a set of colours and $X \subseteq V(G) \cup E(G)$. All colourings considered in this paper are *proper*, that is to say, no two elements of X which are adjacent or incident may be assigned the same colour. In particular, if $X = V(G)$, then φ is called a *vertex-colouring* of G, if $X = E(G)$, then φ is called an *edge-colouring* of G, and if $X = V(G) \cup E(G)$ then φ is called a *total colouring* of G. The least number of colours needed for a vertex-colouring, an edge-colouring, or a total colouring of G, is the *chromatic number* $\chi(G)$, the *chromatic index* $\chi'(G)$, or the *total chromatic number* $\chi''(G)$, respectively.

2. The Conformability Conjecture

A graph G is said to be *conformable* if $V(G)$ can be coloured with $\Delta(G) + 1$ colours in such a way that the number of colour classes having parity different from that of $|V(G)|$ is at most $\operatorname{def}(G)$. Such a colouring of $V(G)$ is called a *conformable vertex-colouring*. When deciding whether a given $(\Delta(G) + 1)$-vertex-colouring is conformable or not, empty colour classes are counted as colour classes of even parity (there are even order graphs for which the only $(\Delta(G) + 1)$-conformable vertex-colouring must include an empty colour class, for example $K_4 - e$.

We show in Section 3 that if $\Delta(G) \geq (|V(G)| + 1)/2$ (a condition the graphs we study here generally satisfy), then there is a polynomial time algorithm to decide whether G is conformable in the case when $|V(G)|$ is even. If $|V(G)|$ is odd and $\Delta(G) \geq (|V(G)| + 1)/2$, it is not clear to us whether such a polynomial algorithm exists.

We first note the following proposition.

PROPOSITION 2.1. *If G is nonconformable, then G is Type 2.*

PROOF. In any total colouring of G with $\Delta(G) + 1$ colours, each colour occurs at each vertex v of degree $\Delta(G)$, either on v itself, or on an edge incident with v. Therefore, if some colour occurs on a number of vertices of parity different from $|V(G)|$, then it contributes at least one to the deficiency of G. It follows that the number of such colours is at most $\operatorname{def}(G)$, in other words, that G is conformable. □

Initial investigations appear to support the idea that nonconformability is the main constraint causing a graph having maximum degree which is large in relation to its order to be Type 2. It is clear that if G contains a subgraph H which has the same maximum degree as G and which is Type 2, then G is Type 2. Chetwynd and Hilton originally conjectured (see [15]) that for a graph with maximum degree more than half its order to be Type 2, it must contain a nonconformable subgraph of the same maximum degree. Chen and Fu (see [9]) showed that the graph obtained by subdividing a single edge of a complete graph of odd order is Type 2, conformable and contains no nonconformable subgraph having the same maximum degree. We will call a graph obtained by subdividing a single edge of a complete graph of odd order, a *Chen and Fu* graph. There is evidence (which will be presented in Theorem 2.5) to suggest that the class of Chen and Fu graphs may be the only class of counterexamples to Chetwynd and Hilton's original conjecture. The revised conjecture that we will now call the Conformability Conjecture states:

CONFORMABILITY CONJECTURE 1. Let G satisfy $\Delta(G) \geq \big(|V(G)| + 1\big)/2$. Then G is Type 2 if and only if either G contains a nonconformable subgraph H with $\Delta(H) = \Delta(G)$, or $\Delta(G)$ is even and G contains a subgraph H obtained by subdividing an edge of $K_{\Delta(G)+1}$.

The lower bound of $\big(|V(G)| + 1\big)/2$ cannot be lowered any further, since $K_{n,n}$ for n even is conformable and Type 2.

In view of the fact that almost all graphs satisfy $\Delta(G) \geq \big(|V(G)|/2 + 1\big)$, the Conformability Conjecture, if true, might seem particularly striking in the light of Sánchez-Arroyo's result [33] that to determine whether or not a graph is Type 1 is NP-hard. However it is not clear whether such a striking dichotomy really exists, since we do not know how hard it is to determine whether or not a graph satisfying $\Delta(G) \geq \big(|V(G)| + 1\big)/2$ is conformable (even more, whether all its subgraphs of the same maximum degree are conformable).

Our definition of Type 2 does not presuppose the truth of the Total Colouring Conjecture of Behzad and Vizing, and so the Conformability Conjecture (as stated here) is independent of it. If, however, we do assume the truth of the Total Colouring Conjecture, then the Conformability Conjecture can be expressed in an alternative form. First let us note the following proposition.

PROPOSITION 2.2. *Let G be a Type 2 graph. Suppose it is true that if H is any graph of maximum degree $\Delta(G) - 1$ with at most $|V(G)|$ vertices, then H satisfies $\chi''(H) \leq \Delta(H) + 2$. Then G contains a critical subgraph of degree $\Delta(G)$.*

PROOF. Remove edges form G until a critical subgraph G^* is obtained such that $\chi''(G^*) = \Delta(G)+2$. If $\Delta(G^*) < \Delta(G)$ then let G^{**} be a supergraph of G^* and a subgraph of G with $\Delta(G^{**}) = \Delta(G)-1$. Then $\chi''(G^{**}) \geq \Delta(G)+2 = \Delta(G^{**})+3$. But this contradicts our assumption. Consequently $\Delta(G^*) = \Delta(G)$. $\qquad\square$

Proposition 2.2 has the following consequences.

COROLLARY 2.3. *Let G be a Type 2 graph satisfying $\Delta(G) \geq 3|V(G)|/4 + 1$. Then G has a critical subgraph of degree $\Delta(G)$.*

PROOF. This follows from the result of Hilton and Hind in [27]. $\qquad\square$

COROLLARY 2.4. *If the Total Colouring Conjecture is true, then each Type 2 graph has a critical subgraph with the same maximum degree.*

It follows from Corollary 2.4 that if we assume the Total Colouring Conjecture, then we can restate the Conformability Conjecture in the following way.

CONFORMABILITY CONJECTURE 2. Let G satisfy $\Delta(G) \geq \big(|V(G)|/2 + 1\big)$. Then G is critical if and only if either G is nonconformable and G contains no proper nonconformable subgraph having maximum degree $\Delta(G)$, or $\Delta(G)$ is even and G is obtained by subdividing an edge of $K_{\Delta(G)+1}$.

There is a further constraint that the vertex-colouring must satisfy if the graph G is to be Type 1. Suppose that G is given a vertex-colouring with $\Delta(G)+1$ colours.

Let the vertex-colour classes be $C_1, C_2, \ldots, C_{\Delta(G)+1}$. For $1 \leq i \leq \Delta(G) + 1$, let ξ_i be the number of vertices whose neighbourhood consists of vertices in C_i. Let

$$\xi_i^+ \begin{cases} = \xi_i & \text{if } \xi_i + |C_i| \equiv |V(G)| \pmod 2, \\ = \xi_i + 1 & \text{if } \xi_i + |C_i| \not\equiv |V(G)| \pmod 2. \end{cases}$$

Then

$$\text{def}(G) \geq \sum_{i=1}^{\Delta(G)+1} \xi_i^+.$$

This is not hard to see. For if all the neighbours of vertices $v_1, v_2, \ldots, v_{\xi_i}$ are coloured with the same c_i, then clearly none of $v_1, v_2, \ldots, v_{\xi_i}$ are coloured c_i and no edge incident with any of $v_1, v_2, \ldots, v_{\xi_i}$ is coloured c_i. Moreover, if $\xi_i + |C_i| \not\equiv |V(G)| \pmod 2$ there must be a further vertex at which c_i is absent. Consequently, the colour class C_i contributes ξ_i^+ to the deficiency.

This stronger but more complicated criterion implies the easier one used in the definition of conformability. It also explains why the Chen and Fu graphs, although conformable, are Type 2. The following theorem shows that the Chen and Fu graphs are unique in the sense that they are the only graphs having fairly large degree that are conformable, but fail to satisfy the more complex condition described above. We have, therefore, retained the current definition of conformability throughout this paper.

Notice that the condition that $\delta(G) \geq 2$, in the statement of the theorem, simply reflects the fact that if v is a vertex of degree one in a graph H, then H has a Type 1 colouring whenever $H \backslash v$ has a Type 1 colouring provided $\Delta(H) \geq 2$. The constant 4 in the condition $\Delta(G) \geq |V(G)|/2 + 4$ can be reduced by introducing a number of extra arguments. It is possible that a bound of the form $\Delta(G) \geq (|V(G)|+1)/2$ could be obtained, although we have not obtained a proof with this bound. Our aim in presenting this result is, however, to indicate that the Chen and Fu graphs have a special nature, and we feel that the bound on $\Delta(G)$ illustrates this while limiting the length of the proof.

THEOREM 2.5. *Let G be a graph having $\delta(G) \geq 2$ and $\Delta(G) \geq |V(G)|/2 + 4$. Let G be conformable. Then G has a vertex colouring for which*

$$\text{def}(G) \geq \sum_{i=1}^{\Delta+1} \xi_i^+$$

if and only if G is not a Chen and Fu graph.

PROOF. Let $\Delta = \Delta(G)$. Given a vertex-colouring φ of G with $\Delta + 1$ colours, let

$$f(\varphi) = \sum_{i=1}^{\Delta+1} \xi_i^+ - \text{def}(G).$$

NECESSITY. Suppose that G is a Chen and Fu graph. Let $|V(G)| = 2n$. There are essentially three distinct ways of colouring $V(G)$ with $\Delta+1 = 2n-1$ colours. Let x denote the vertex of degree 2. In one possible vertex colouring, there is an empty colour class, two doubleton colour classes, and $\Delta - 2$ singleton colour classes; one doubleton colour class consists of the two neighbours of x, and the other doubleton colour class contains x. The other two colourings are obtained by recolouring one vertex of a doubleton colour class with the colour of the empty colour class. The latter two vertex-colourings are nonconformable. The former is conformable but satisfies $f(\varphi) = 1$.

SUFFICIENCY. For $1 \leq i \leq \Delta$, let p_i be the number of vertices of degree i in G. We begin by noting that $|V(G)| - 1 \geq \Delta \geq |V(G)|/2 + 4$, so $|V(G)| \geq 10$ and $\Delta \geq 9$.

For given $|V(G)|$ and Δ, choose φ to be a conformable vertex-colouring for which $f(\varphi)$ is minimal, and, subject to this, for which the number of singleton colour classes is a minimum, and subject to both these conditions, for which the number of empty colour classes is a minimum. We shall assume that G is not a Chen and Fu graph. The result will be proved if we can show that $f(\varphi) \leq 0$. We shall in fact assume that $f(\varphi) \geq 1$ and, in every case, obtain a contradiction.

Let X be the set of vertices $x \in V(G)$ such that $x \in X$ if and only if the neighbourhood $N(x)$ of x is contained in a single colour class; the colour classes containing the neighbourhoods may be different for different vertices in X. Clearly

$$|X| = \sum_{i=1}^{\Delta+1} \xi_i.$$

If $|X| = 0$, then $\xi_i = 0$ ($1 \leq i \leq \Delta + 1$) so $\sum_{i=1}^{\Delta+1} \xi_i^+$ is the number of colour classes with parity different from $|V(G)|$; since φ is conformable, this number is at most $\operatorname{def}(G)$, so $f(\varphi) \leq 0$, a contradiction. Therefore $|X| \geq 1$.

Let r be the number of even colour classes.

Now suppose that $|V(G)| = 2n+1$ for some integer n. If φ has no empty colour classes, then the largest colour class contains at most $(2n + 1) - (\Delta + (r - 1)) = 2n - \Delta - r + 2$ vertices. Therefore, for each $x \in X$, $d_G(x) \leq 2n - \Delta - r + 2$. Consequently

$$\operatorname{def}(G) \geq \sum_{x \in X} \big(\Delta - d_G(x)\big) \geq |X|(2\Delta + r - 2n - 2) \geq |X|(r+6).$$

Since $\sum_{i=1}^{\Delta+1} \xi_i = |X|$ and $\xi_i + |C_i| \not\equiv (2n+1) \pmod 2$ for at most $r + |X|$ values of i, it follows that

$$\sum_{i=1}^{\Delta+1} \xi_i^+ \leq |X| + \big(r + |X|\big) = 2|X| + r < 6|X| + r|X| \leq \operatorname{def}(G),$$

so $f(\varphi) \leq 0$, a contradiction. Now suppose that φ has an empty colour class, say $C_{\Delta+1} = \emptyset$. We may suppose that $N(x) \subseteq C_1$ for some $x \in X$. Choose $v \in N(x)$ and let $C_1' = C_1 \setminus \{v\}$ and $C_{\Delta+1}' = \{v\}$. Let φ' be the resulting vertex-colouring, where the remaining colour classes are unchanged. In φ' there are no more colour classes which have parity different from $|V(G)|$ than there are in φ, so φ' is also conformable. Moreover $\xi_1'^+ \leq \xi_1^+$ and $0 = \xi_{\Delta+1}'^+ < \xi_{\Delta+1}^+ = 1$, so $f(\varphi') < f(\varphi)$, a contradiction. Therefore $|V(G)| \neq 2n + 1$. Thus $|V(G)| = 2n$ for some integer n.

Since $|V(G)|$ is even, $\sum_{i=1}^{\Delta+1} \xi_i = |X|$ and $\xi_i + |C_i| \not\equiv (2n) \pmod 2$ for at most $\Delta + 1 - r + |X|$ values of i, it follows that

$$\sum_{i=1}^{\Delta+1} \xi_i^+ \leq 2|X| + \Delta + 1 - r.$$

We show that G contains at most one vertex of degree 2. Suppose there exists a colour class C_i for which $|C_i| \geq \Delta - 1$. Since no vertex in C_i is adjacent to any other vertex in C_i,

$$\begin{aligned} \mathrm{def}(G) &\geq |C_i|\big(\Delta - (2n - |C_i|)\big) \\ &\geq (\Delta - 1)(2\Delta - 1 - 2n) \\ &\geq (n+3)(7) = 7n + 21. \end{aligned}$$

Using the very crude bounds $|X| \leq 2n$ and $\Delta \leq 2n - 1$, we have

$$\sum_{i=1}^{\Delta+1} \xi_i^+ \leq 2|X| + \Delta + 1 - r \leq 6n - r,$$

so $f(\varphi) = \sum_{i=1}^{\Delta+1} \xi_i^+ - \mathrm{def}(G) \leq 0$, a contradiction. Thus we may assume that for each $x \in X$, $|N(x)| \leq |C_i| \leq \Delta - 2$, for some i. Note that each vertex of degree 2 contributes $(\Delta - 2)$ to the deficiency, so

$$\mathrm{def}(G) \geq \big(|X| - p_2\big)2 + p_2(\Delta - 2) = 2|X| + p_2\Delta - 4p_2.$$

It follows that

$$\begin{aligned} 1 \leq f(\varphi) &\leq \big(2|X| + \Delta + 1 - r\big) - \big(2|X| + p_2\Delta - 4p_2\big) \\ &= (1 - p_2)(\Delta - 4) + 1 + 4 - r, \end{aligned}$$

so that $(p_2 - 1)(\Delta - 4) \leq 4$. Since $\Delta \geq 9$, it follows that $p_2 \leq 1$.

We now consider two possible cases.

CASE 1. φ has no empty colour classes.

In this case the largest colour class contains at most $2n - \Delta \leq n - 4 \leq \Delta - 8$ colours. It follows that if $x \in X$ then $d_G(x) \leq \Delta - 8$, so in particular $d_G(x) \leq \Delta - 2$.

Suppose there are two vertices $x_1, x_2 \in X$ whose neighbourhoods are contained in two distinct colour classes, say $N(x_1) \subseteq C_1$ and $N(x_2) \subseteq C_2$. Then

$$\begin{aligned} \mathrm{def}(G) &\geq \sum_{x \in X} \big(\Delta - d_G(x)\big) \geq \big(\Delta - d_G(x_1)\big) + \big(\Delta - d_G(x_2)\big) + \big(|X| - 2\big)2 \\ &\geq 2\Delta - |C_1| - |C_2| + 2|X| - 4, \end{aligned}$$

so

$$\begin{aligned} 1 \leq f(\varphi) &\leq 2|X| + \Delta + 1 - r - \big(2\Delta - |C_1| - |C_2| + 2|X| - 4\big) \\ &\leq -\Delta + 5 + |C_1| + |C_2|, \end{aligned}$$

so that $|C_1| + |C_2| \geq \Delta - 4$. Since each colour class contains at least one vertex,

$$\Delta - 1 \leq \sum_{i=3}^{\Delta+1} |C_i| = 2n - |C_1| - |C_2| \leq 2n - \Delta + 4,$$

so $\Delta < n + 4$, a contradiction. Therefore the neighbours of each vertex $x \in X$ are contained in the same colour class, say C_1.

Let v_j be a vertex in a singleton colour class C_j. Suppose that there is a $k \notin \{1, j\}$ such that v_j is not adjacent to any vertex of C_k. Let $C_k' = C_k \cup \{v_j\}$ and $C_j' = \emptyset$, and let φ' be the vertex-colouring obtained, where the remaining colour classes are unchanged. Then φ' is conformable. If C_k' does not contain the neighbourhood of any vertex in the complement of X then $f(\varphi') \leq f(\varphi)$, a contradiction, since φ' has fewer singleton colour classes than φ. So there is a vertex $y \notin X$ such that $N(y) \subseteq C'_k$. But then, if $x \in X$,

$$\text{def}(G) \geq \big(\Delta - d_G(y)\big) + \sum_{x \in X} \big(\Delta - d_G(x)\big)$$
$$\geq 2\Delta - d_G(x) - d_G(y) + 2\big(|X| - 1\big)$$
$$\geq 2\Delta - |C_1| - |C_k| + 2|X| - 2,$$

and we obtain a contradiction just as just above (with C_k instead of C_2). It follows that each vertex in a singleton colour class is adjacent to at least one vertex in every other colour class, apart from C_1 possibly.

Since there are $\Delta + 1$ colour classes, all nonempty, each vertex in a singleton colour class is adjacent to at most one vertex in C_1.

Since $\Delta \geq n+4$ and φ has no empty colour classes, if s is the number of singleton colour classes, then $2n \geq 2(\Delta+1-s)+s$ so $s \geq 2\Delta+2-2n \geq 2(n+4)+2-2n = 10$. Since $p_2 \leq 1$, there is a vertex in a singleton colour class that is nonadjacent to any vertex of degree 2. We may suppose that $v_{\Delta+1}$ is such a vertex in the colour class $C_{\Delta+1}$.

Let $x \in X$. Since $N(x) \subseteq C_1$ and $d_G(x) \geq 2$, $|C_1| \geq 2$. Therefore there is a $z \in N(x)$ such that $zv_{\Delta+1} \notin E(G)$. Let $C'_1 = C_1 \backslash \{z\}$ and $C'_{\Delta+1} = \{v_{\Delta+1}, z\}$, and let φ' be the vertex-colouring obtained, the remaining colour classes being unchanged. Then φ' is conformable. But since $v_{\Delta+1}$ is not adjacent to any vertex of degree 2, $C'_{\Delta+1}$ does not contain the neighbourhood of any vertex, and so $f(\varphi') < f(\varphi)$, a contradiction.

CASE 2. φ has at least one empty colour class; say $C_{\Delta+1} = \emptyset$.

Suppose $|C_i| \geq 4$ for some $i \in \{1, 2, \ldots, \Delta+1\}$. Since $p_2 \leq 1$, there are vertices $v, v' \in C_i$ which are not both neighbours of a vertex of degree two. Let $C'_i = C_i \backslash \{v, v'\}$ and $C'_{\Delta+1} = \{v, v'\}$, and let φ' be the vertex-colouring obtained, where the remaining colour classes are unchanged. Then φ' is conformable. Moreover, since $C'_{\Delta+1}$ does not contain the neighbourhood of any vertex, $f(\varphi') \leq f(\varphi)$. But φ' has the same number of singleton colour classes as φ, and one fewer empty colour classes, a contradiction. Therefore $|C_i| \leq 3$ for $1 \leq i \leq \Delta+1$.

It follows that only vertices of degree two or three can be in X, so $p_2+p_3 \geq |X|$. Then, recalling that $\sum \xi_i^+ \leq 2|X| + \Delta + 1 - r$, and that $r \geq 1$ since $C_{\Delta+1} = \emptyset$,

$$1 \leq f(\varphi) = \sum_{i=1}^{\Delta+1} \xi_i^+ - \text{def}(G)$$
$$\leq (2p_2 + 2p_3 + \Delta) - \big(p_2(\Delta - 2) + p_3(\Delta - 3)\big)$$
$$= \Delta - p_2(\Delta - 4) - p_3(\Delta - 5).$$

Recall that $\Delta \geq 9$. The inequality implies that $p_2+p_3 \leq 1$, except possibly if $\Delta = 9$ and $p_2 = 0$, $p_3 = 2$. In this exceptional case, $|V(G)| = 10$, and then the deficiency cannot just be concentrated at the two vertices of degree three, and we find that $\text{def}(G) \geq 14$. Therefore, in the exceptional case, $1 \leq f(\varphi) \leq (2 \cdot 0 + 2 \cdot 2 + 9) - 14 = -1$, a contradiction. Consequently, in each case, $p_2 + p_3 \leq 1$, so G has at most one vertex, say x, of degree at most 3. Moreover, $1 \leq |X| \leq p_2 + p_3 \leq 1$, so $|X| = 1$ and so $X = \{x\}$. We may suppose that $N(x) \subseteq C_1$.

If $|C_1| = 3$, then $\xi_1^+ = \xi_1 = 1$. Since φ is conformable, φ has at most $\text{def}(G)-1$ other odd colour classes, so $f(\varphi) = \sum_{i=1}^{\Delta+1} \xi_i^+ - \text{def}(G) \leq 1 + \big(\text{def}(G)-1\big) - \text{def}(G) =$

0, a contradiction. Therefore $|C_1| < 3$. Since $d_G(x) \geq 2$ and $N(x) \subseteq C_1$, it follows that $d_G(x) = 2$ and that $|C_1| = 2$.

We now have that

$$1 \leq f(\varphi) = \sum_{i=1}^{\Delta+1} \xi_i^+ - \operatorname{def}(G) \leq \left(2|X| + \Delta + 1 - r\right) - (\Delta - 2) = 5 - r,$$

so that $r \leq 4$. Since φ has at least one empty colour class, $1 \leq r \leq 4$. Thus G has at most three doubleton colour classes.

If φ had no singleton colour classes, then since $r \leq 4$, φ would have at least $\Delta - 3$ colour classes of size at least 3, so $(\Delta - 3)3 \leq 2n$. Therefore $n + 4 \leq \Delta \leq (2/3)n + 3$, which is impossible. Thus φ has at least one singleton colour class.

Let w be a vertex in a singleton colour class. Since $d_G(x) = 2$, $p_2 = 1$ and $N(x) = C_1$, the only vertex of degree 2 has its neighbourhood in C_1. So w must be adjacent to every other vertex contained in a singleton colour class, for otherwise we could combine the two singleton colour classes producing a conformable vertex-colouring with $f(\varphi)$ reduced, a contradiction. Also w must be adjacent to every vertex of each colour class containing three vertices, for otherwise we could replace the three vertices and w with two doubleton colour classes, and obtain a conformable vertex-colouring with $f(\varphi)$ reduced, a contradiction. Furthermore, w must be adjacent to at least one vertex in every doubleton colour class C_i with $i \neq 1$, for otherwise we could add w to C_i; this would produce a conformable vertex-colouring with $f(\varphi)$ the same, but with the number of singleton colour classes reduced. Finally, w must be adjacent to both vertices in the doubleton colour class C_1, since if w were not adjacent to a vertex in C_1, we could remove that vertex from C_1 and add it to the colour class containing w, thereby reducing $f(\varphi)$, a contradiction. Since G has at most three doubleton colour classes it follows that w is nonadjacent to at most two vertices, so $\Delta \geq 2n - 3$; moreover, if $\Delta = 2n - 3$ then φ has exactly three doubleton colour classes.

Now suppose that $\Delta = 2n - 3$. Since G contains a vertex of degree two, G must also contain at least one further vertex with even degree. Consequently $\operatorname{def}(G) \geq (\Delta - 2) + 1 = \Delta - 1$. Therefore

$$1 \leq f(\varphi) = \sum_{i=1}^{\Delta+1} \xi_i^+ - \operatorname{def}(G) \leq \left(2|X| + \Delta + 1 - r\right) - (\Delta - 1) = 4 - r,$$

so $1 \leq r \leq 3$. Therefore φ has at most two doubleton colour classes, a contradiction. Consequently $\Delta \geq 2n - 2$.

Suppose that φ has a colour class, say C_i, of order 3. Let C_j be a singleton colour class containing w, say. If $x \in C_i$, then, as $N(x) \subseteq C_1$, x is nonadjacent to w. Consequently, if we put $C_i' = C_i \setminus \{x\}$ and $C_j' = \{w, x\}$, we obtain a conformable vertex-colouring with a reduced value of $f(\varphi)$, a contradiction. So suppose that $x \notin C_i$. Each vertex in C_i is nonadjacent to the other two vertices in C_i and so has degree at most $2n - 3 \leq \Delta - 1$. Therefore $\operatorname{def}(G) \geq (\Delta - 2) + 3 = \Delta + 1$, and so

$$1 \leq f(\varphi) \leq \sum_{i=1}^{\Delta+1} \xi_i^+ - \operatorname{def}(G) \leq \left(2|X| + \Delta + 1 - r\right) - (\Delta + 1) = 2 - r.$$

Since φ has an empty colour class and a doubleton colour class, namely C_1, this is a contradiction. Therefore φ contains only empty, singleton and doubleton colour classes.

Since $N(x) = C_1$ and x can't be in a singleton colour class, as it is nonadjacent to all vertices which are in singleton colour classes, φ has at least two doubleton colour classes. Therefore $r \geq 3$.

If $\Delta = 2n - 1$, then G contains a vertex of degree two, namely x, and at least two further vertices (those in C_1) of degree at most $2n - 2 = \Delta - 1$. Consequently $\mathrm{def}(G) \geq (\Delta - 2) + 2 = \Delta$. Therefore

$$1 \leq f(\varphi) \leq \sum_{i=1}^{\Delta+1} \xi_i^+ \leq \left(2|X| + \Delta + 1 - r\right) - (\Delta) = 3 - r,$$

so $1 \leq r \leq 2$, a contradiction.

If $\Delta = 2n - 2$ then G is contained in the Chen and Fu graph on $2n$ vertices. Since, by assumption G is not the Chen and Fu graph, G is a proper subgraph of it, and so $\mathrm{def}(G) \geq \Delta$. Consequently, if φ is the conformable vertex-colouring of the Chen and Fu graph described in the proof of the necessity, then, with respect to G, $f(\varphi) \leq 0$, a contradiction. $\qquad\square$

In the following two sections, we will present results characterising graphs which are critical and which have large maximum degree. While doing so a number of results will also be shown to support the Conformability Conjecture.

3. Conformability and Criticality for Even Order Graphs

From a practical point of view, when $\Delta(G) \geq \left(|V(G)|+1\right)/2$ the nonconformability condition takes on a different form depending on whether $|V(G)|$ is odd or even. In the case when $|V(G)|$ is even, there is a simple inequality determining when G is conformable, and there is a polynomial algorithm to decide whether this inequality is satisfied, and thus to decide whether G is conformable. If the Conformability Conjecture 1B is correct, then this results in a simple test to decide whether G is critical. We also demonstrate the rather surprising fact that, if Conjecture 1B is correct, and if $\left(|V(G)|+1\right)/2 \leq \Delta(G) \leq 2|V(G)|/3 - 1$, then G is critical if and only if G is regular and has a rather special structure.

Before beginning our detailed study, we introduce some notation. We use $\overline{G}$ to denote the complement of G. We use $e(G)$ to denote $|E(G)|$, $j(G)$ to denote the edge independence number of G, and $o(G)$ to denote the number of odd order components of G.

We shall need Tutte's One-Factor Theorem in the more structural form due to Berge.

LEMMA 3.1 (Berge [**31**]). *If H is a graph having order p and edge independence number j, then*

$$p - 2j = \max_{S \subseteq V(H)} \left\{o(H - S) - |S|\right\}.$$

We begin this section by giving two results that establish the Conformability Conjecture for graphs having very large maximum degree.

THEOREM 3.2. *Let G be a graph of even order $2n \geq 6$ and maximum degree $\Delta = 2n - 1$. Then G is critical if and only if either $e(\overline{G}) + j(\overline{G}) = n - 1$, or $e(\overline{G}) + j(\overline{G}) = n - 2$ and $\overline{G}$ consists of a number of disjoint complete graphs of odd order. Also K_2 and K_4 are critical.*

PROOF. SUFFICIENCY. Hilton [24] showed that if G is a graph of even order $2n \geq 6$ and maximum degree $\Delta = 2n - 1$, then G is Type 2 if and only if $e(\overline{G}) + j(\overline{G}) \leq n - 1$. Therefore if $e(\overline{G}) + j(\overline{G}) = n - 1$ then G is Type 2, and, for any edge e of G, $e(\overline{G} \cup \{e\}) + j(\overline{G} \cup \{e\}) \geq n$, so $G \backslash \{e\}$ is Type 1. Thus G is critical. Now suppose that $e(\overline{G}) + j(\overline{G}) = n - 2$ and $\overline{G}$ consists of a number of disjoint complete graphs of odd order. If e is any edge of G then $e(\overline{G} \cup \{e\}) = e(\overline{G}) + 1$ and $j(\overline{G} \cup \{e\}) = j(\overline{G}) + 1$ so that $e(\overline{G} \cup \{e\}) + j(\overline{G} \cup \{e\}) \geq n$, so $G \backslash \{e\}$ is Type 1. Thus again G is critical.

NECESSITY. If G is critical, then G is Type 2 so $e(\overline{G}) + j(\overline{G}) \leq n - 1$. If $e(\overline{G}) + j(\overline{G}) \leq n - 3$ and e is any edge of G, then $e(\overline{G} \cup \{e\}) + j(\overline{G} \cup \{e\}) \leq (n-3) + 2 = n - 1$, so $G \backslash \{e\}$ is Type 2, so G is not critical. Therefore $n - 2 \leq e(\overline{G}) + j(\overline{G}) \leq n - 1$.

Suppose that $e(\overline{G}) + j(\overline{G}) = n - 2$ and that G is critical. Then by Hilton's result [24], for each edge e of G, $j(\overline{G} \cup \{e\}) = j(\overline{G}) + 1$. By Lemma 3.1, there is some set $S \subseteq V(G)$ such that

$$2j(\overline{G}) = 2n - o(\overline{G} - S) + |S|.$$

In any set of $j(\overline{G})$ independent edges of $\overline{G}$, there are $|S|$ of them joining vertices in S to vertices in the odd components of $\overline{G} - S$. If $|S| \neq 0$ and e is an edge of G joining a vertex of degree zero in $\overline{G}$ to a vertex in S, then $j(\overline{G} \cup \{e\}) = j(\overline{G})$, a contradiction. Therefore $|S| = 0$. If $\overline{G}$ contains an even component C, then $j(C) = |V(C)|/2$, and so, if e is an edge of G joining a vertex of C to a vertex of degree zero in $\overline{G}$, then $j(C \cup \{e\}) = j(C)$, so $j(\overline{G} \cup \{e\}) = j(\overline{G})$, a contradiction. Therefore $\overline{G}$ contains no even components. If C is an odd component of $\overline{G}$ and $|V(C)| \geq 3$, then $j(C) = (|V(C)| - 1)/2$. If C is not a complete graph, then there is an edge e of G such that $V(C \cup \{e\}) = V(C)$, and then $j(\overline{G} \cup \{e\}) = j(\overline{G})$, a contradiction. Thus $\overline{G}$ consists of disjoint complete graphs of odd order. □

Note that for a graph G having order $2n$ and maximum degree $2n - 1$, Theorem 3.4 shows that the condition $e(\overline{G}) + j(\overline{G}) \leq n - 1$ implies that G is nonconformable.

PROPOSITION 3.3 (Chen and Fu, [9]). *Let G be a graph of even order $2n \geq 4$ and maximum degree $\Delta = 2n - 2$. Then G is critical if and only if G is K_{2n-1} with one edge subdivided.*

In the following theorem we show that, for graphs with even order and maximum degree at least half the order, nonconformability is equivalent to an inequality involving simple graph parameters. The case when $\Delta = 2n - 1$ was proved by Chetwynd and Hilton in [15]. The general form of the inequality was first considered by Yap [38] where he proved part (a) of the Theorem.

THEOREM 3.4. *Let G be a graph of even order $2n$.*
(a) *If $e(\overline{G}) + j(\overline{G}) \leq n(2n - \Delta) - 1$ then G is nonconformable.*
(b) *If G is nonconformable and has maximum degree $\Delta \geq n - 1$, then $e(\overline{G}) + j(\overline{G}) \leq n(2n - \Delta) - 1$.*

PROOF. We first notice the equivalence of the following inequalities

$$e(\overline{G}) + j(\overline{G}) \geq n(2n - \Delta)$$
$$j(\overline{G}) \geq \tfrac{1}{2}\big(2n + 2n(2n - \Delta - 1) - 2e(\overline{G})\big)$$
$$j(\overline{G}) \geq \tfrac{1}{2}\big(2n - \mathrm{def}(G)\big)$$
$$\mathrm{def}(G) \geq 2n - 2j(\overline{G}).$$

If G has a conformable vertex-colouring, then $V(G)$ can be coloured with $\Delta + 1$ colours so that at most $\mathrm{def}(G)$ colours are assigned to an odd number of vertices. Therefore there is a set of at least $2n - \mathrm{def}(G)$ vertices upon which each colour occurs an even number of times. This implies that $2j(\overline{G}) \geq 2n - \mathrm{def}(G)$ and thus, from the equivalence of inequalities shown above, that $e(\overline{G}) + j(\overline{G}) \geq n(2n - \Delta)$. This establishes part (a) of the proof.

Conversely, suppose that $e(\overline{G}) + j(\overline{G}) \geq n(2n - \Delta)$ and that $\Delta \geq n - 1$. We colour $j(\overline{G})$ disjoint independent pairs of vertices with $j(\overline{G})$ colours, and colour the remaining $2n - 2j(\overline{G})$ vertices with a further $2n - 2j(\overline{G})$ colours, thus using at most $2n - j(\overline{G})$ colours. A result of Erdős and Pósa [19] states that

$$j(\overline{G}) \geq \min\{\tfrac{1}{2}|V(\overline{G})|, \delta(\overline{G})\}$$
$$= \min\{n, 2n - \Delta - 1\}.$$

Since $\Delta \geq n-1$, it follows that $2n - \Delta - 1 \leq n$ and therefore that $j(\overline{G}) \geq 2n - \Delta - 1$. Thus $\Delta + 1 \geq 2n - j(\overline{G})$ and we have used at most $\Delta + 1$ colours to colour the vertices of G. Since $\mathrm{def}(G) \geq 2n - 2j(\overline{G})$, at most $\mathrm{def}(G)$ colours are assigned to an odd number of vertices. The vertex-colouring just described is therefore a conformable vertex-colouring for G. This establishes part (b). □

Next we deduce a structural form of the Conformability Conjecture for graphs having even order and maximum degree between half and three quarters of their order. We will need the following lemma:

LEMMA 3.5. *Let H be a graph with deficiency $\mathrm{def}(H)$. Then $\mathrm{def}(H)$ is odd if and only if both $\Delta(H)$ and $|V(H)|$ are odd.*

PROOF. By definition

$$\mathrm{def}(H) = \sum_{v \in V(H)} \big(\Delta - d_H(v)\big)$$
$$= |V(H)|\Delta(H) - \sum_{v \in V(H)} d_H(v)$$
$$= |V(H)|\Delta(H) - 2|E(H)|$$
$$\equiv |V(H)|\Delta(H) \pmod 2. \quad \square$$

We state the following theorem in terms of a parameter ℓ in order to emphasize that similar results for larger values of ℓ are obtainable; the descriptions of the nonconformable graphs simply become more complex.

THEOREM 3.6. *Let $\ell = 1$ or 2 and let G be a graph having even order $2n$. If*

$$\frac{\ell}{\ell + 1}(2n) - 1 < \Delta \leq \frac{\ell + 1}{\ell + 2}(2n) - 1,$$

then either G is conformable or regular. If G is not conformable and $\ell = 1$, then G contains a K_{r_1,r_2}, where $r_1 + r_2 = 2n$, r_1, r_2 are odd and $r_1, r_2 \geq 2n - \Delta$. If G is not conformable and $\ell = 2$, then either G contains a K_{r_1,r_2} as above or G contains a K_{r_1,r_2,r_3}, where $r_1 + r_2 + r_3 = 2n - 1$, r_1, r_2, r_3 are odd, $r_1, r_2 \geq 2n - \Delta$ and $r_3 \geq 2n - \Delta - 1$.

PROOF. Suppose G is nonconformable. If $\ell = 1$, then the bounds on Δ imply that $n \geq 3$, so $2n \geq 6 = (\ell+1)(\ell+2)$. If $\ell = 2$ and $n \in \{1, 2, 3, 4, 5\}$, then the only possible options for which there exist integral values of Δ satisfying the bounds are $n = 2$ and $\Delta = 2$ (in which case no such nonconformable graph exists), $n = 4$ and $\Delta = 5$ (in which case $G = \overline{K_3 \cup C_5}$ which is Graph 14 in the catalogue of critical graphs in Section 8) or $n = 5$ and $\Delta = 6$ (again, no such graph exists; to see this, note that it follows from Theorem 3.4 that any such graph must be regular of degree 3. But any 3-regular graph of order 10 has a 1-factor, so it follows from Theorem 3.4 that there is no such graph). So if $\ell = 2$ and $2n \leq 10$ the result holds; thus we may assume that $2n \geq 12 = (\ell+1)(\ell+2)$ in this case too. We therefore assume throughout the remainder of the proof that $2n \geq (\ell+1)(\ell+2)$.

By Theorem 3.4 and the equivalence of the inequalities $e(\overline{G}) + j(\overline{G}) < n(2n - \Delta)$ and $\operatorname{def}(G) < 2n - 2j(\overline{G})$, there is a number $k > 0$ such that

$$j(\overline{G}) = n - \frac{1}{2}\operatorname{def}(G) - k.$$

By Lemma 3.5, $\operatorname{def}(G)$ is even, and so k is a positive integer. By Lemma 3.1, there is a vertex-set $S \subseteq V(G)$ such that

$$o(\overline{G} - S) = |S| + \operatorname{def}(G) + 2k.$$

Let $s = |S|$.

We make the following definitions: A component of $\overline{G} - S$ will be called *large* if the order of the component is greater than $\overline{\delta} = \delta(\overline{G})$. Let c^* denote the number of large components of $\overline{G} - S$ and let c_o^* be the number of large odd components of $\overline{G} - S$. A component of $\overline{G} - S$ will be called *small* if the order of the component is between 1 and $\overline{\delta}$, inclusive. For $1 \leq i \leq \overline{\delta}$, let c_i denote the number of components of order i in $\overline{G} - S$. Then

$$o(\overline{G} - S) = \sum_{i=1, i \text{ odd}}^{\overline{\delta}} c_i + c_o^*.$$

Since $\overline{\delta} = 2n - 1 - \Delta$ and $\Delta \leq \{(\ell+1)/(\ell+2)\}2n - 1$ it follows that $(\ell+2)\overline{\delta} \geq (\ell+2)(2n-1) - (\ell+1)2n + \ell + 2 = 2n$, so that

$$\overline{\delta} \geq \frac{1}{\ell+2}(2n).$$

Since each large component contains at least $(\overline{\delta} + 1)$ vertices,

$$c^* \leq \left\lfloor \frac{2n}{\overline{\delta}+1} \right\rfloor \leq (\ell+2) - 1 = \ell + 1.$$

Note that as $2n \geq (\ell+1)(\ell+2)$ it follows that $\overline{\delta} \geq 2n/(\ell+2) \geq \ell+1$, so $\overline{\delta} \geq c^*$.

We now consider two main cases:

CASE 1. $c^* = \ell + 1$.

If there is at least one small component of order i say, then there are at least $\overline{\delta} + 1 - i$ vertices in S and so $|V(\overline{G})| \geq c^*(\overline{\delta}+1) + i + (\overline{\delta}+1-i) = (\ell+2)(\overline{\delta}+1) \geq$

$2n + \ell + 2$, a contradiction. Therefore if $c^* = \ell + 1$ then $\overline{G} - S$ contains no small components.

One consequence of this is that

$$c_o^* = o(\overline{G} - S) = s + \mathrm{def}(G) + 2k \geq s + \mathrm{def}(G) + 2.$$

If $c_o^* = 2$, then it follows that $s = \mathrm{def}(G) = 0$, so G is regular. If $\ell = 1$ then $\overline{G}$ contains no even components, so G contains a K_{r_1, r_2} where r_1, r_2 are odd, $r_1 + r_2 = 2n$ and $r_1, r_2 \geq 2n - \Delta$. If $\ell = 2$ then $\overline{G}$ contains one large even component, so G contains a K_{r_1, r_2, r_3} where r_1, r_2, r_3 are odd, $r_1 + r_2 + r_3 = 2n - 1$, $r_1, r_2 \geq \overline{\delta} + 1 \geq 2n - \Delta$ and $r_3 \geq (\overline{\delta} + 1) - 1 \geq 2n - \Delta - 1$.

If $c_o^* \neq 2$, then, since $3 \geq \ell + 1 = c^* > 2$, it follows that $c_o^* = 3$. Then $3 \geq s + \mathrm{def}(G) + 2$, and so, as $\mathrm{def}(G)$ is even, it again follows that $\mathrm{def}(G) = 0$, and thus that G is regular. Moreover, as three large odd components contain an odd number of vertices altogether, $s \neq 0$, so $s = 1$. In this case G again contains a K_{r_1, r_2, r_3} where r_1, r_2, r_3 are odd, $r_1 + r_2 + r_3 = 2n - 1$, $r_1, r_2, r_3 \geq \overline{\delta} + 1 \geq 2n - \Delta$.

CASE 2. $c^* \leq \ell$.

Let ζ be the number of edges in $E(\overline{G})$ that are incident with a vertex in S and with a vertex in $\overline{G} - S$. The sum $\sum_{v \in S} d_{\overline{G}}(v)$ is not more than $\overline{\delta} s + \mathrm{def}(G)$. Consequently,

$$\zeta \leq \overline{\delta} s + \mathrm{def}(G).$$

If C is a small odd component of $\overline{G} - S$ with order i, then at least $i(\overline{\delta} + 1 - i)$ edges join vertices in C to vertices in S. Consequently

$$\zeta \geq \sum_{\substack{i=1, i \text{ odd}}}^{\overline{\delta}} i(\overline{\delta} + 1 - i)c_i.$$

Notice that $i(\overline{\delta} + 1 - i) - \overline{\delta} = (i - 1)(\overline{\delta} - i) \geq 0$ for $1 \leq i \leq \overline{\delta}$, so that $i(\overline{\delta} + 1 - i) \geq \overline{\delta}$ with equality only if $i = 1$ or $i = \overline{\delta}$.

It therefore follows that

$$\overline{\delta} s + \mathrm{def}(G) \geq \zeta \geq \sum_{\substack{i=1, i \text{ odd}}}^{\overline{\delta}} i(\overline{\delta} + 1 - i)c_i \geq \overline{\delta} \sum_{\substack{i=1, i \text{ odd}}}^{\overline{\delta}} c_i$$

$$= \overline{\delta}(o(\overline{G} - S) - c_o^*)$$

$$= \overline{\delta}(s + \mathrm{def}(G) + 2k - c_o^*).$$

Since $c_o^* \leq c^* \leq \ell \leq 2$, $\overline{\delta} \geq \ell + 1 \geq 2$ and $2k \geq 2$, it follows that

$$\mathrm{def}(G) = 0, \quad k = 1$$

and

$$c_o^* = c^* = \ell = 2.$$

Thus G is also regular in this case and $\overline{G} - S$ has two large odd components and no large even component.

It remains to show that G contains either a 2-partite or a 3-partite complete subgraph as described in the statement of the theorem. To do this we first show that $\overline{G} - S$ contains no component of order less than $\overline{\delta}$. In the arguments that follow we will frequently use the fact that $\overline{\delta} = 2n - \Delta - 1 \geq (2n)/4$ to derive a contradiction.

Since we have shown that $\overline{\delta}s + \mathrm{def}(G) = \overline{\delta}\big(s + \mathrm{def}(G) + 2k - c_o^*\big)$, it follows that each intermediate inequality derived while obtaining this result must also be an equality. In particular,

$$\sum_{i=1, i \text{ odd}}^{\overline{\delta}} c_i(i)(\overline{\delta} + 1 - i) = \overline{\delta} \sum_{i=1, i \text{ odd}}^{\overline{\delta}} c_i.$$

Recall that this equality can only hold provided that $c_i = 0$ for every $1 < i < \overline{\delta}$.

Suppose that $c_1 \neq 0$. Then at least $\overline{\delta}$ edges join each singleton component of $\overline{G} - S$ (that is, each component comprising a single vertex) to the vertices of S, so $s \geq \overline{\delta}$. It follows that $c_{\overline{\delta}} = 0$ for otherwise $\overline{G}$ would contain at least $s + c_1 + c_{\overline{\delta}}\overline{\delta} + c_0^*(\overline{\delta} + 1) \geq \overline{\delta} + 1 + \overline{\delta} + 2(\overline{\delta} + 1) \geq 4\overline{\delta} + 3 > 2n$ vertices, a contradiction. Therefore $c_1 = \sum_{i=1, i \text{ odd}}^{\overline{\delta}} c_i = s + \mathrm{def}(G) + 2k - c_o^* = s$. But since $c_1 = s$ and $s \geq \overline{\delta}$, it follows that $\overline{G}$ contains at least $s + c_1 + c_0^*(\overline{\delta} + 1) \geq \overline{\delta} + \overline{\delta} + 2(\overline{\delta} + 1) \geq 4\overline{\delta} + 2 > 2n$ vertices, a contradiction. Therefore we must conclude that $c_1 = 0$.

Thus $c_i = 0$ for every $1 \leq i < \overline{\delta}$. Since $\sum_{i=1, i \text{ odd}}^{\overline{\delta}} c_i = s$ it follows that, if $\overline{\delta}$ is odd, then $c_{\overline{\delta}} = s$, and if $\overline{\delta}$ is even then $s = 0$.

If $c_{\overline{\delta}} \neq 0$, then $c_{\overline{\delta}} = 1$, for otherwise $\overline{G} - S$ would contain at least $2\overline{\delta} + 2(\overline{\delta} + 1) = 4\overline{\delta} + 2 \geq 2n + 2$ vertices, a contradiction. Therefore $s = c_{\overline{\delta}} = 1$ and $\overline{\delta}$ is odd. Since $s = 1$, each component of $\overline{G} - S$ contains at least $\overline{\delta}$ vertices. Since $\overline{\delta}$ is odd, any even component of $\overline{G} - S$ has at least $\overline{\delta} + 1$ vertices, and so is large, a contradiction. Therefore $\overline{G} - S$ has no even components. Consequently $\overline{G} - S$ contains three odd components, two having order at least $\overline{\delta} + 1$, the other having order $\overline{\delta}$; also $s = 1$. Therefore G contains a K_{r_1, r_2, r_3} where r_1, r_2, r_3 are odd, $r_1 + r_2 + r_3 = 2n - 1$, $r_3 \geq \overline{\delta} \geq 2n - \Delta - 1$ and $r_1, r_2 \geq \overline{\delta} + 1 \geq 2n - \Delta$.

It remains to consider the case when $c_{\overline{\delta}} = 0$. In this case $s = 0$. Any even component of $\overline{G} - S$ would therefore have to contain at least $\overline{\delta} + 1$ vertices, and so be a large component. Since we have already shown that $\overline{G} - S$ has no large even components, it follows that $\overline{G}$ consists of exactly two large odd components. Therefore G contains a K_{r_1, r_2} where r_1, r_2 are odd, $r_1 + r_2 = 2n$ and $r_1, r_2 \geq \overline{\delta} + 1 \geq 2n - \Delta$. $\square$

Note that the Conformability Conjecture 1 and Theorem 3.6 imply the following:

CONFORMABILITY CONJECTURE 3. Let G be a graph having even order $2n$ and maximum degree $n + 1 \leq \Delta \leq 3/2n - 1$. Then G is critical if and only if G is regular and G contains either a K_{r_1, r_2}, where r_1, r_2 are odd, $r_1 + r_2 = 2n$ and $r_1, r_2 \geq 2n - \Delta$ or a K_{r_1, r_2, r_3}, where r_1, r_2, r_3 are odd, $r_1 + r_2 + r_3 = 2n - 1$, r_1, $r_2 \geq 2n - \Delta$ and $r_3 \geq 2n - \Delta - 1$.

We show in Theorem 6.4 that, in one special case, the graphs described in Theorem 3.6 are critical. These are the regular graphs obtained from $K_{n+2, n-2}$ by adding the square of an $(n+2)$-cycle to the independent vertex subset of size $n + 2$.

It is worth commenting that although further classes of graphs which are almost certainly critical can be produced from complete r-partite graphs for higher values of Δ, regularity ceases to be a requirement if $\Delta \geq 3(2n)/4$. Furthermore, classes of nonconformable graphs, which do not contain a "large" complete r-partite subgraph, exist for these larger values of Δ. To see this consider the graph obtained when $n \equiv 2 \pmod 4$ by taking two copies of $\overline{K}_{n/2}$ and two copies of $K_{n/2}$, joining each vertex which lies in either $\overline{K}_{n/2}$ to all the vertices in the other $\overline{K}_{n/2}$ and to all the vertices in both the $K_{n/2}$'s, and adding a matching of size $n/2$ whose edges join vertices in the two $K_{n/2}$'s. The resulting graph is nonconformable, regular, with order $2n$ and degree $(3/4)(2n)$ but does not contain a K_{r_1,r_2,r_3,r_4} where $r_1 + r_2 + r_3 + r_4 > n + 4$; it is almost certainly critical.

As the final part of this section, we show that the Conformability Conjecture 2 in the case when $|V(G)|$ is even is equivalent to the following Conjecture 4. The equivalence is far from obvious. However, if Conjecture 4 is correct, then whether or not a graph G is critical is determined solely by whether a simple equation involving $e(\overline{G})$, $j(\overline{G})$, $\Delta(G)$ and $|V(G)|$ is satisfied, and there is no need to examine any subgraphs.

CONFORMABILITY CONJECTURE 4. *Let G be a graph with $|V(G)| = 2n$ and $\Delta(G) \geq n + 1$. Then G is critical if and only if either*

  (i) $e(\overline{G}) + j(\overline{G}) = n(2n - \Delta(G))$, $\Delta(G) = 2n - 2$ *and G is $K_{\Delta(G)+1}$ with one edge subdivided, or*
  (ii) $e(\overline{G}) + j(\overline{G}) = n(2n - \Delta(G)) - 1$, *or*
  (iii) $e(\overline{G}) + j(\overline{G}) = n(2n - \Delta(G)) - 2$, *and $\overline{G}$ consists of a number of disjoint complete graphs of odd order.*

Of course, in view of Theorem 3.6, when $3|V(G)|/4 - 1 \geq \Delta(G) \geq |V(G)|/2 + 1$, it is quite easy to demonstrate the equivalence of Conjecture 4 and Conjecture 3.

If true Conjecture 4 would provide a very clear criterion to decide whether a graph is critical, and since there is a polynomial algorithm (for example, see Chapter 9 of [**31**]) to determine $j(\overline{G})$, there would be a polynomial algorithm to determine whether a graph were critical. The criterion in Conjecture 2 does not have this easy aspect, so it is of considerable interest that Conjecture 2 for $|V(G)|$ even, and Conjecture 4, are equivalent.

THEOREM 3.7. *Conjecture 4 is true if and only if Conjecture 2 is true for even order graphs.*

Theorem 3.7 follows from Theorems 3.9 and 3.10.

We first prove a lemma showing that if a graph has reasonably large maximum degree and reasonably small deficiency, then the removal of any two or more vertices leaves a subgraph which has smaller maximum degree or is conformable or is one of four exceptional graphs. Define J_1 to be the graph $K_7 + \overline{K}_3 + K_2$ where the $+$ indicates the join operation, so that each vertex of the $\overline{K}_3$ is joined to each vertex of the K_7 and of the K_2, but no vertex of the K_7 is joined to any vertex of the K_2. Define J_2 to be the graph $K_5 + \overline{K}_{1,2} \tilde{+} K_2$ where the $\tilde{+}$ indicates that the two vertices of the K_2 are joined to the isolated vertex in the $\overline{K}_{1,2}$ and to distinct vertices of degree one in the $\overline{K}_{1,2}$. Let J_3 be the graph $K_4 + \overline{K}_2 +' K_2$ where the $+'$ indicates that each vertex of the K_2 is joined to a distinct vertex of the $\overline{K}_2$. Finally, let J_4 be the graph $K_4 + \overline{K}_2 +'' K_2$ where the $+''$ indicates that one vertex of the K_2 is joined to both of the vertices of the $\overline{K}_2$ and the other is not joined to either vertex of the $\overline{K}_2$.

LEMMA 3.8. *Let G be a graph having maximum degree $\Delta \geq (2/3)|V(G)|$ and deficiency $\mathrm{def}(G) \leq \Delta + 2$. Then G does not have a nonconformable subgraph H with $\Delta(H) = \Delta$ and $|V(H)| \leq |V(G)| - 2$, unless $|V(H)| = |V(G)| - 2$ and G is one of J_1, J_2, J_3 and J_4.*

PROOF. It is sufficient to show that either G is one of the four exceptional graphs or that G has no induced subgraph H with $\Delta(H) = \Delta$ and $|V(H)| \leq |V(G)| - 2$ that is nonconformable.

Suppose that such an induced nonconformable subgraph H exists. Let $W = V(G)\backslash V(H)$ and, by counting the minimum number of edges which join W to the remainder of the graph, notice that

$$(*) \qquad \operatorname{def}(H) \geq |W|(\Delta - |W| + 1) - \operatorname{def}(G).$$

CASE 1. $|V(H)| < |V(G)| - 2$.

In this case $|W| \geq 3$. Since $\Delta \geq 2|V(G)|/3$ and $\Delta \leq |V(H)| - 1 \leq (|V(G)| - 3) - 1$, it follows that $\Delta \geq 8$. If $\Delta = 8$, then $|V(G)| = 12$ and thus $|V(H)| = 9$. For this value of Δ, since $|V(H)|$ is odd, assigning a distinct colour to each vertex in $V(H)$ gives a conformable vertex-colouring for H. Thus we may assume that $\Delta \geq 9$.

Let $f(w) = w(\Delta - w + 1)$. It is easy to see that $f'(w)$ is positive, and thus that $f(w)$ is increasing, for $0 \leq w < (\Delta + 1)/2$. Furthermore, since $(2/3)|V(G)| \leq \Delta$ and $\Delta \leq |V(H)| - 1 = (|V(G)| - |W|) - 1$, it follows that

$$|W| \leq |V(G)| - \Delta - 1 \leq |V(G)| - \tfrac{2}{3}|V(G)| - 1 \leq \tfrac{\Delta}{2} - 1.$$

Thus $f(w)$ is increasing for all allowable values of $w = |W|$ and so $f(|W|) \geq f(3) = 3(\Delta - 2)$. In particular this means that

$$\operatorname{def}(H) - (\Delta + 1) \geq f(|W|) - \operatorname{def}(G) - (\Delta + 1)$$
$$\geq 3(\Delta - 2) - (\Delta + 2) - (\Delta + 1)$$
$$= \Delta - 9 \geq 0.$$

Thus H must be conformable, contradicting the original assumption.

CASE 2. $|V(H)| = |V(G)| - 2$.

In this case $|W| = 2$. Let $W = \{w_1, w_2\}$. Since $\Delta \geq 2|V(G)|/3$ and $\Delta \leq |V(H)| - 1 = |V(G)| - 3$, it follows that $\Delta \geq 6$. Furthermore, since $\operatorname{def}(G) \leq \Delta + 2$,

$$\operatorname{def}(H) \geq 2(\Delta - 2 + 1) - (\Delta + 2) \geq \Delta - 4.$$

If $|V(G)|$ is odd then $|V(H)|$ is odd. Let φ be a vertex colouring of H with no empty colour classes. Let p be the number of colour classes containing two or more vertices. Then there are $\Delta + 1 - p$ singleton colour classes. So

$$(\Delta + 1 - p) + 2p \leq |V(H)| = |V(G)| - 2 \leq \tfrac{3}{2}\Delta - 2.$$

Recalling that $\Delta \geq 6$, it follows that $p \leq \Delta/2 - 3 < \Delta - 4 \leq \operatorname{def}(H)$, so that the number of even colour classes is at most $\operatorname{def}(H)$, a contradiction since H is nonconformable.

Thus we may assume that $|V(G)|$ and $|V(H)|$ are even. In this case we need to take a little more care estimating $\operatorname{def}(H)$. In particular, let

$$\alpha = \sum_{v \in V(G)\backslash W} (\Delta - d_G(v)).$$

Then the number of edges of G between W and $V(H)$ is

$$\begin{cases} 2\Delta - 2 - (\operatorname{def}(G) - \alpha) & \text{if } w_1 w_2 \in E(G), \\ 2\Delta - (\operatorname{def}(G) - \alpha) & \text{if } w_1 w_2 \notin E(G). \end{cases}$$

Therefore

$$\text{def}(H) = \begin{cases} \alpha + 2\Delta - 2 - \big(\text{def}(G) - \alpha\big) & \text{if } w_1 w_2 \in E(G), \\ \alpha + 2\Delta - \big(\text{def}(G) - \alpha\big) & \text{if } w_1 w_2 \notin E(G). \end{cases}$$

Since $\text{def}(G) \leq \Delta + 2$ it follows that

$$\text{def}(H) \geq \begin{cases} \Delta - 4 + 2\alpha & \text{if } w_1 w_2 \in E(G), \\ \Delta - 2 + 2\alpha & \text{if } w_1 w_2 \notin E(G). \end{cases}$$

Note that $\alpha \geq 0$, so that $\text{def}(H) > 0$ as $\Delta \geq 6$. We also have that

$$\text{def}(G) = \begin{cases} 2\alpha + 2\Delta - 2 - \text{def}(H) & \text{if } w_1 w_2 \in E(G), \\ 2\alpha + 2\Delta - \text{def}(H) & \text{if } w_1 w_2 \notin E(G). \end{cases}$$

CASE 2.1. $\Delta = \big|V(H)\big| - 1$

Since Δ is odd and $\text{def}(G)$ and $\text{def}(H)$ are both even, it follows that

$$\Delta + 1 \geq \text{def}(G) \quad \text{and} \quad \text{def}(H) \geq \begin{cases} \Delta - 3 + 2\alpha & \text{if } w_1 w_2 \in E(G), \\ \Delta - 1 + 2\alpha & \text{if } w_1 w_2 \notin E(G). \end{cases}$$

Since H is nonconformable, the number of odd colour classes in any colouring of $V(H)$ with $\Delta + 1$ colours is greater than $\text{def}(H)$. Hence $\Delta + 1 > \text{def}(H)$. Since $\Delta + 1$ and $\text{def}(H)$ are both even, $\Delta - 1 \geq \text{def}(H)$. Therefore $\alpha \leq 1$. Since $\text{def}(H) > 0$, H is not complete, so H has a vertex colouring with two nonadjacent vertices receiving one colour, and the remaining $\big|V(H)\big| - 2 = \Delta - 1$ vertices receiving distinct colours. If $\alpha = 1$, or if $\alpha = 0$ and $w_1 w_2 \notin E(G)$, or if $\alpha = 0$, $w_1 w_2 \in E(G)$ and $\text{def}(H) \geq \Delta - 1$, then this vertex colouring is conformable, a contradiction. Therefore we only need to consider the possibility that $\alpha = 0$, $w_1 w_2 \in E(G)$, or $\alpha = 0$, w_1, w_2 are elements of $E(G)$ and $\text{def}(H) = \Delta - 3$. In this case $\text{def}(G) = \Delta + 1$.

The graph H cannot have two distinct pairs of nonadjacent vertices, for otherwise the vertex-colouring giving two colours to the vertices in these pairs and a different colour to the remaining $\Delta - 3 = \text{def}(H)$ vertices is conformable, a contradiction. Therefore $\overline{H} = K_3 \cup r K_1$ for some positive integer r, or $\overline{H} = K_{1,s} \cup t K_1$ for some positive integers s and t.

If $\overline{H} = K_3 \cup r K_1$, then $\text{def}(H) = 6 = \Delta - 3$, so $\Delta = 9$. Recalling that $\alpha = 0$ it follows that w_1 and w_2 are both adjacent to each of the vertices in the K_3 in $\overline{H}$. Since $w_1 w_2 \in E(G)$ it follows that $G = J_1$.

If $\overline{H} = K_{1,s} \cup t K_1$, then either $s = 1$ or $s \geq 2$. If $s \geq 2$ then since $\alpha = 0$, the hub vertex of the star is joined in G to $\Delta = \big|V(H)\big| - 1 = \big|V(G)\big| - 3$ vertices, and so actually $s = 2$. Then $\text{def}(H) = 4 = \Delta - 3$, so $\Delta = 7$. It follows that $G = J_2$.

If $s = 1$ then $\text{def}(H) = 2 = \Delta - 3$, so $\Delta = 5$, and each vertex of the $\overline{K_2}$ is joined to a vertex of W. Therefore G is J_3 or J_4.

CASE 2.2. $\Delta \leq \big|V(H)\big| - 2$.

Firstly, notice that Szemerédi and Hajnal's result [22] guarantees that H has an equitable vertex-colouring using $\Delta + 1$ colours. Since $\Delta \geq 2\big|V(G)\big|/3$, it follows that H has a vertex-colouring having only singleton and doubleton vertex colour classes. Suppose φ is a vertex-colouring having p_0 empty, p_1 singleton and p_2 doubleton vertex colour classes (and no larger colour classes), where $p_2 = j(\overline{H})$. Then

$$p_1 + 2p_2 = \big|V(H)\big|, \quad p_1 + p_2 \leq \Delta + 1,$$

so that $p_2 \geq |V(H)| - \Delta - 1$.

Since H is nonconformable and $\mathrm{def}(H) \geq \Delta - 4 + 2\alpha$, it follows that $p_1 \geq \Delta - 3 + 2\alpha$. Consequently $p_2 \leq \Delta + 1 - p_1 \leq 4 - 2\alpha$. Since $p_2 \leq 4 - 2\alpha$,

$$4 - 2\alpha \geq |V(H)| - \Delta - 1$$

so

$$\Delta \geq |V(H)| - 5 + 2\alpha.$$

Therefore

$$|V(H)| - 5 + 2\alpha \leq \Delta \leq |V(H)| - 2,$$

so again $0 \leq \alpha \leq 1$. Since $p_1 \geq \Delta - 3 + 2\alpha$ and $p_2 = (|V(H)| - p_1)/2$,

$$p_2 \leq \tfrac{1}{2}(|V(H)| - \Delta + 3 - 2\alpha).$$

Since $p_2 = j(\overline{H})$, writing $\Delta = |V(H)| - c$, we have

$$j(\overline{H}) \leq \tfrac{1}{2}(c + 3 - 2\alpha).$$

Secondly, notice that if $\Delta = |V(H)| - c$ and $\Delta \geq 2|V(G)|/3 = 2(|V(H)| + 2)/3$, then $|V(H)| \geq 4 + 3c$.

Thirdly, notice that if $v \in V(H)$, then

$$d_H(v) \geq \Delta - \alpha - 2$$

since at most two edges of G join v to vertices in W. Thus

$$\Delta(\overline{H}) = |V(H)| - 1 - \delta(H) \leq |V(H)| - 1 - \Delta + \alpha + 2 = c + 1 + \alpha.$$

Since $\Delta(H) = \Delta$,

$$\delta(\overline{H}) = |V(H)| - 1 - \Delta(H) = c - 1.$$

By Lemma 3.1 there exists a set S for which

$$o(\overline{H} - S) - |S| = |V(\overline{H})| - 2j(\overline{H}).$$

Let t_i be the number of components of $\overline{H} - S$ having order i. Every maximum matching in $\overline{H}$ must include edges joining each vertex in S to a distinct odd component of $\overline{H} - S$. A component of $\overline{H} - S$ of order i contains $\lfloor i/2 \rfloor$ independent edges. Thus

$$j(\overline{H}) = |S| + \sum_{i \geq 2} \lfloor \tfrac{i}{2} \rfloor t_i.$$

Therefore

$$\tfrac{1}{2}(c + 3 - 2\alpha) \geq j(\overline{H}) \geq |S| + \sum_{i \geq 2} \lfloor \tfrac{i}{2} \rfloor t_i.$$

Therefore

$$\tfrac{3}{2}(c + 3 - 2\alpha) - 3|S| - 3t_2 \geq 3 \sum_{i \geq 3} \lfloor \tfrac{i}{2} \rfloor t_i \geq \sum_{i \geq 3} i t_i,$$

so at most $3(c + 3 - 2\alpha)/2 - 3|S| - 3t_2$ vertices are in components having order 3 or more. Therefore

$$2t_2 + t_1 + |S| \geq |V(H)| - \left\{ \tfrac{3}{2}(c + 3 - 2\alpha) - 3|S| - 3t_2 \right\}$$
$$\geq 4 + 3c - \tfrac{3}{2}c - \tfrac{9}{2} + 3\alpha + 3|S| + 3t_2.$$

Therefore

$$t_1 \geq -\tfrac{1}{2} + \tfrac{3}{2}c + 2|S| + 3\alpha.$$

Each edge leaving a singleton odd component of $\overline{H} - S$ must be incident with a new vertex in S, and so

$$t_1 \delta(\overline{H}) \leq \Delta(\overline{H})|S|.$$

But since $\delta(\overline{H}) = c - 1$, we have

$$t_1 \delta(\overline{H}) \geq \left(-\tfrac{1}{2} + \tfrac{3}{2}c + 2|S| + 3\alpha\right)(c - 1).$$

The inequality

$$\left(-\tfrac{1}{2} + \tfrac{3}{2}c + 2|S| + 3\alpha\right)(c - 1) > (c + 1 + \alpha)|S|$$

may be demonstrated as follows: If $c \geq 4$ then

$$\left(-\tfrac{1}{2} + \tfrac{3}{2}c + 2|S| + 3\alpha\right)(c - 1) > 2(c - 1)|S| \geq (c + 2)|S| \geq (c + 1 + \alpha)|S|,$$

since $\alpha \in \{0, 1\}$.

If $c = 3$ then $|S| \leq j(\overline{H}) \leq \tfrac{1}{2}(c + 3 - 2\alpha) \leq 3$, so

$$\left(-\tfrac{1}{2} + \tfrac{3}{2}c + 2|S| + 3\alpha\right)(c - 1) \geq \left(4 + 2|S|\right)2$$
$$= 8 + 4|S| > 5|S| = (c + 2)|S| \geq (c + 1 + \alpha)|S|.$$

If $c = 2$, then $|S| \leq (c + 3 - 2\alpha)/2 = 5/2 - \alpha$, so $|S| \leq 2$. Therefore

$$\left(-\tfrac{1}{2} + \tfrac{3}{2}c + 2|S| + 3\alpha\right)(c - 1) = \tfrac{5}{2} + 2|S| + 3\alpha > (3 + \alpha)|S| = (c + 1 + \alpha)|S|.$$

Consequently we have

$$t_1 \delta(\overline{H}) > (c + 1 + \alpha)|S|$$

and since $(c + 1 + \alpha) \geq \Delta(\overline{H})$ we obtain finally that

$$t_1 \delta(\overline{H}) > \Delta(\overline{H})|S|,$$

a contradiction. Thus no such nonconformable graph exists.

Thus we conclude that if $|V(G)|$ is even and $|W| = 2$, then the only possible graphs are J_1, J_2, J_3 or J_4. □

We now show that if Conjecture 2 is true for graphs of even order, then Conjecture 4 is true. In other words, we prove the following result.

THEOREM 3.9. *Let G be a graph of even order $2n \geq 6$ and maximum degree $\Delta \geq n + 1$. If the Conformability Conjecture 2 is correct for graphs of even order, then G is critical if and only if one of the following is true:*

(i) *$e(\overline{G}) + j(\overline{G}) = n(2n - \Delta)$, $\Delta = 2n - 2$ and G is $K_{\Delta+1}$ with one edge subdivided, or*

(ii) *$e(\overline{G}) + j(\overline{G}) = n(2n - \Delta) - 1$, or*

(iii) *$e(\overline{G}) + j(\overline{G}) = n(2n - \Delta) - 2$ and $\overline{G}$ consists of a number of disjoint complete graphs of odd order.*

PROOF. If $\Delta = 2n - 1$ or $2n - 2$ then the theorem can be shown to be true without assuming Conjecture 1. If $\Delta = 2n - 1$ it follows by Theorem 3.2. Chen and Fu [9] actually showed that the only Type 2 graph of order $2n$ with $\Delta = 2n - 2$ is the graph $K_{\Delta+1}$ with one edge subdivided (this statement is equivalent to

Proposition 3.3), so, by Proposition 2.1 and Theorem 3.4, cases (ii) and (iii) cannot occur. The theorem in the case $\Delta = 2n - 2$ follows from this.

From now on we shall assume that $\Delta \leq 2n - 3$ and that Conjecture 2 is correct for graphs of even order.

NECESSITY. (for $n + 1 \leq \Delta \leq 2n - 3$).

Assume that (i), (ii) or (iii) holds. Then in fact one of (ii) or (iii) holds. Let e be an arbitrary edge of G and let $G^* = G \backslash e$. By the Erdős-Pósa lemma [20], $j(\overline{G}) \geq \delta(\overline{G}) = 2n - \Delta - 1$, so

$$\sum_{v \in V(G)} d_{\overline{G}}(v) = 2e(\overline{G}) \leq 2n(2n - \Delta) - 2 - j(\overline{G})$$

$$\leq 2n(2n - \Delta) - 2 - 2(2n - \Delta - 1)$$
$$= (2n - 2)(2n - \Delta).$$

Therefore

$$\sum_{v \in V(G)} d_G(v) \geq 2n(2n - 1) - (2n - 2)(2n - \Delta)$$

$$= 2n(\Delta - 1) + 2(2n - \Delta) \geq 2n(\Delta - 1) + 6.$$

It follows that $\Delta(G^*) = \Delta(G) = \Delta$. We also have that

$$e(\overline{G^*}) + j(\overline{G^*}) = n(2n - \Delta) + c$$

where $c \in \{0, 1\}$. If $c = 1$ then case (ii) applied and $j(\overline{G^*}) = j(\overline{G}) + 1$.

By Theorem 3.4 the graph G^* is conformable. The necessity will follow from Conjecture 2 for even order graphs if we show that G^* contains no nonconformable subgraphs of the same maximum degree; for then it follows that G^* is Type 1, so G is critical.

CASE 1. $(4/3)n \leq \Delta \leq 2n - 3$.

Suppose that G^* has a nonconformable subgraph H. It follows as in the argument at the start of the proof of Theorem 3.4 that

$$\operatorname{def}(G^*) = 2n - 2j(\overline{G^*}) + 2c.$$

By the Erdős-Posá result[20],

$$j(\overline{G^*}) \geq \min\{n, 2n - \Delta - 1\} = 2n - \Delta - 1,$$

so it follows that

$$\operatorname{def}(G^*) \leq 2n - 2(2n - \Delta - 1) + 2c$$

with equality only if $j(\overline{G^*}) = 2n - \Delta - 1$. Therefore

$$\operatorname{def}(G^*) \leq \Delta - (2n - \Delta) + 2 + 2c \leq \Delta + 2,$$

so by Lemma 3.8,

$$H = G \backslash w \quad \text{for some } w \in V(G^*).$$

Let φ be a vertex-colouring of H with $\Delta + 1$ colours with no empty colour classes. Such a colouring exists since $|V(H)| = 2n - 1 \geq \Delta + 1$ and we can split colour classes if necessary. Let p_i be the number of colour classes of φ with cardinality i. Then

$$\sum_{i \geq 1} p_i = \Delta + 1 \quad \text{and} \quad \sum_{i \geq 1} i p_i = 2n - 1,$$

so the number of even colour classes is

$$\sum_{i \text{ even}} p_i \leq \sum_{i \geq 1} (i-1) p_i = 2n - \Delta - 2.$$

The inequality $(*)$ in the proof of Lemma 3.8 applies with $|W| = 1$, so

$$\begin{aligned}
\operatorname{def}(H) &\geq \Delta - \operatorname{def}(G^*) \\
&\geq \Delta - \big(2n - 2(2n - \Delta - 1) + 2c\big) \\
&= 2n - \Delta - 2 - 2c,
\end{aligned}$$

with equality only if $j(\overline{G^*}) = 2n - \Delta - 1$. It follows that the number of even colour classes possessed by φ is at most $\operatorname{def}(H)$, yielding the contradiction that φ is a conformable vertex-colouring of H, except possibly if $c = 1$ and $j(\overline{G^*}) = 2n - \Delta - 1$. However, by the Erdős-Posá result [20] again,

$$j(\overline{G}) \geq \min\{n, 2n - \Delta - 1\} = 2n - \Delta - 1.$$

Thus $j(\overline{G^*}) > j(\overline{G}) \geq 2n - \Delta - 1 = j(\overline{G^*})$, a contradiction.

Therefore all subgraphs H of G^* with $\Delta(H) = \Delta(G)$ are conformable in this case.

CASE 2. $n + 1 \leq \Delta < (4/3)n$.

Since $(4/3)n > n + 1$ it follows that $n \geq 4$ so $\Delta \geq n + 1 \geq 5$. Furthermore, since G is nonconformable, by Theorem 3.6 G is regular, so $\operatorname{def}(G^*) = 2$. Let H be a subgraph of G^*. Let $W = V(G^*) \backslash V(H)$.

CASE 2a. $|W| \geq 2$.

In this case, as with $(*)$ at the start of the proof of Lemma 3.8,

$$\operatorname{def}(H) \geq |W|(\Delta - |W| + 1) - \operatorname{def}(G^*).$$

Since $\Delta \geq n + 1$, it follows that $|V(G)| = 2n \leq 2(\Delta - 1)$. Therefore $|W| = |V(G)| - |V(H)| \leq 2(\Delta - 1) - (\Delta + 1) = \Delta - 3$, so that

$$2 \leq |W| \leq \Delta - 3.$$

From the expression above for $\operatorname{def}(H)$ it follows that

$$\begin{aligned}
\operatorname{def}(H) &\geq \min\big\{2(\Delta - 2 + 1) - 2, (\Delta - 3)(\Delta - (\Delta - 3) + 1) - 2\big\} \\
&= \min\{2\Delta - 4, 4\Delta - 14\} \geq \Delta + 1,
\end{aligned}$$

since $\Delta \geq 5$. Therefore H is conformable in this case.

CASE 2b. $|W| = 1$.

Since $\delta(G^*) = \Delta - 1$ and at least one vertex not in W has degree $\Delta - 1$ in G^*, it follows that

$$\operatorname{def}(H) \geq (\Delta - 1) + 1 = \Delta.$$

Since $|V(H)|$ is odd, any vertex-colouring of H with $\Delta + 1$ colours has at least one odd colour class and so is conformable.

SUFFICIENCY. (for $n + 1 \leq \Delta \leq 2n - 3$).

Let G be a critical graph. Then G is nonconformable. By Theorem 3.4, since $\Delta \geq n + 1$,

$$e(\overline{G}) + j(\overline{G}) \leq n(2n - \Delta) - 1.$$

If $e(\overline{G}) + j(\overline{G}) \leq n(2n - \Delta) - 3$ and e is any edge of G, then $e(\overline{G} \cup \{e\}) + j(\overline{G} \cup \{e\}) \leq n(2n - \Delta) - 1$, so by Theorem 3.4, $G \backslash e$ is nonconformable, and so $G \backslash e$ is Type 2. This contradicts the assumption that G is critical. Thus

$$n(2n - \Delta) - 2 \leq e(\overline{G}) + j(\overline{G}) \leq n(2n - \Delta) - 1.$$

In order to complete the proof of the sufficiency, we must consider further the case

$$e(\overline{G}) + j(\overline{G}) = n(2n - \Delta) - 2.$$

In this case it is clear that for each edge e of G, $j(\overline{G} \cup \{e\}) = j(\overline{G}) + 1$. By Lemma 3.1, there is some set $S \subseteq V(G)$ such that

$$2j(\overline{G}) = 2n - o(\overline{G} - S) + |S|.$$

In any set of $j(\overline{G})$ independent edges in $\overline{G}$, there are $|S|$ of them joining vertices in $|S|$ to vertices in the odd components of $\overline{G} - S$. Thus any set of $j(\overline{G})$ independent edges consists of $|S|$ edges joining distinct points of $|S|$ to vertices in distinct odd components, and each odd component of order, say x, contributes $(x-1)/2$ independent edges.

If any vertex of S is joined in G by an edge e to any other vertex in any component of $\overline{G} - S$, then $j(\overline{G} \cup \{e\}) = j(\overline{G})$, a contradiction. Therefore, in $\overline{G}$, each vertex in S is joined to each vertex not in S. Similarly, if any two vertices in any of the components of $\overline{G} - S$ are joined in G by an edge e, then $j(\overline{G} \cup \{e\}) = j(\overline{G})$, a contradiction. Therefore, in $\overline{G} - S$, each of these components is a complete graph. Again, if any vertex in any even component of $\overline{G} - S$ is joined in G to any vertex in any odd component of $\overline{G} - S$ by an edge e, then $j(\overline{G} \cup \{e\}) = j(\overline{G})$, a contradiction. Hence there is no even component. Finally, if any vertex in S is joined in G to any other vertex in S by an edge e, then $j(\overline{G} \cup \{e\}) = j(\overline{G})$, a contradiction. Therefore, in $\overline{G}$, each vertex in S is joined to every other vertex in S.

The above arguments show that each vertex in S is adjacent in $\overline{G}$ to every other vertex in $\overline{G}$. So each vertex in S would be an isolated vertex in G. Since G is critical, and thus connected, it follows that S is the empty set. Since we have seen that the components of $\overline{G} - S = \overline{G}$ are complete subgraphs of odd order, the sufficiency follows.   □

Finally, we show the converse to Theorem 3.9, namely that if Conjecture 4 is true then Conjecture 2 is true for even order graphs. In other words, we prove the following result.

THEOREM 3.10. *Let G be a graph of even order $2n \geq 6$ and maximum degree $\Delta \geq n + 1$. If the Conformability Conjecture 4 is correct, then G is critical if and only if either G is nonconformable and G contains no proper nonconformable subgraph having maximum degree Δ, or Δ is even and G is obtained by subdividing an edge of $K_{\Delta+1}$.*

PROOF. Suppose that the Conformability Conjecture 4 is true. Suppose that G is critical and that G is not obtained by subdividing an edge of $K_{\Delta+1}$. Then (ii) or (iii) of Conjecture 4 is true. Then, by Theorem 3.4(a), G is nonconformable, and, as G is critical, G cannot contain a proper nonconformable subgraph of the same maximum degree.

Now suppose that G is nonconformable and that G contains no proper nonconformable subgraph of the same maximum degree Δ, or that G is obtained by subdividing an edge of $K_{\Delta+1}$. In the latter case, (i) of Conjecture 4 applies, so G is critical. Consider the former case. Then by Theorem 3.4, $e(\overline{G}) + j(\overline{G}) = n(2n - \Delta) - 1$ or $e(\overline{G}) + j(\overline{G}) = n(2n - \Delta) - 2$ and the removal of any edge from G increases the edge independence number j. As in the proof of the necessity in Theorem 3.9, this implies that $\overline{G}$ consists of a number of disjoint complete graphs of odd order. Thus (ii) or (iii) of Conjecture 4 are satisfied. Therefore, by Conjecture 4, G is critical. $\qquad\square$

4. Conformability and Criticality for Odd Order Graphs

The main results in this section, Theorem 4.15 and Corollary 4.16, are quite substantial verifications of the Conformability Conjecture 2 for graphs of odd order.

Regrettably, however, there does not appear to be an analogue of Theorem 3.4 for graphs of odd order. Because of this we have been unable so far to say whether there is a polynomial algorithm to determine whether or not an odd order graph G with $\Delta(G) \geq (1/2)\big(|V(G)| + 1\big)$ is conformable. Another consequence is that, even assuming Conjecture 2, we have been unable to give a satisfactory description of critical graphs of odd order and high maximum degree; however we do have a result in this direction in Theorems 4.18.

As in the previous section, we begin by presenting some results for odd order graphs having very large maximum degree which lend support to the Conformability Conjecture.

Since it is well known that K_{2n+1} is Type 1, the following result holds:

PROPOSITION 4.1. *Let G be a graph of odd order $2n + 1 \geq 3$ and maximum degree $\Delta = 2n$. Then G is Type 1 and is thus not critical.*

The next result is a consequence of work by Chen, Fu and Yap [10].

PROPOSITION 4.2 (Chen, Fu and Yap [10]). *Let G be a graph of odd order $2n + 1 \geq 3$ and maximum degree $\Delta = 2n - 1$. Then G is not critical.*

It is easy to see that no such graph can be regular. Thus a vertex-colouring assigning one colour to a pair of nonadjacent vertices and each of the $\Delta = 2n - 1$ remaining colours to a unique vertex is a conformable vertex-colouring. This shows that every graph satisfying the conditions of Proposition 4.2 is conformable.

Before proceeding, we need to introduce a number of lemmas. By Vizing's Theorem [34], the edge-chromatic number is bounded by $\Delta(G) \leq \chi'(G) \leq \Delta(G)+1$. We shall call a graph Class 1 if $\chi'(G) = \Delta(G)$ and Class 2 if $\chi'(G) = \Delta(G) + 1$. The first lemma is due to Chetwynd and Hilton [13].

LEMMA 4.3 (Chetwynd and Hilton [13]). *Let G be a connected graph with exactly three vertices of maximum degree. Then G is Class 2 if and only if G has three vertices of degree $|V(G)| - 1$ and the remainder have degree $|V(G)| - 2$. This condition implies that $|V(G)|$ is odd.*

The next lemma is also due to Chetwynd and Hilton and is proved in [13].

LEMMA 4.4 (Chetwynd and Hilton [13]). *Let G be a regular graph of order $2n$ and degree $d(G) = 2n - 1$, $2n - 2$, $2n - 3$, $2n - 4$ or $2n - 5$. If $d(G) \geq |V(G)|/2$, then G is Class 1.*

The next lemma is a simple consequence of Dirac's Theorem giving conditions for the existence of a hamiltonian cycle; an extension due to Berge [5] of a theorem of Chvátal gives a stronger condition, but we will not need this extra strength in this paper.

LEMMA 4.5. *Let G be a graph having even order p and minimum degree δ. Let E be an independent subset of $E(G)$ and F be any subset of $E(G)$ satisfying $E \cap F = \emptyset$. If*

$$\delta \geq \tfrac{1}{2}p + |E| + \Delta_F,$$

where Δ_F is the maximum degree of the subgraph of G induced by the edge set F, then G contains a perfect matching including all the edges of E and excluding all the edges of F.

The next result shows that critical graphs having odd order do exist when $\Delta = |V| - 3$. In our proof of this theorem, for graphs of order ≤ 9 we refer to our catalogue of critical graphs of order ≤ 10, instead of providing a formal proof.

THEOREM 4.6. *Let G be a regular graph of odd order $2n + 1$ and degree $d = 2n - 2$. Then G is critical if and only if $\overline{G}$ does not contain a K_3, i.e., if and only if G is nonconformable.*

PROOF. Since G is regular, $\text{def}(G) = 0$, so G is conformable if and only if $V(G)$ can be coloured by $\Delta + 1$ colours in such a way that each colour class is odd. Since $2n + 1 = (2n - 2) \cdot 1 + 1 \cdot 3$, this is possible if and only if G contains an independent set of three vertices, i.e., $\overline{G}$ contains a K_3.

We shall show initially that G is Type 2 if and only if $\overline{G}$ does not contain a K_3.

If $\overline{G}$ does not contain a K_3, then G is nonconformable and so is Type 2. Conversely, suppose that $\overline{G}$ contains a K_3. Then we notice that $n \neq 2$, for if $n = 2$, then $\overline{G} = C_5$. If $n = 3$, then $|V(G)| = 7$ and then $\overline{G}$ comprises a K_3 and a disjoint C_4 and a total colouring of G using five colours is easily found. Therefore suppose that $n \geq 4$. Let a, b, c be three vertices forming a K_3 in $\overline{G}$. Let $H = G \backslash \{a, b, c\}$. Then H has order $2n - 2$ and minimum degree at least $2n - 5 \geq (2n - 2)/2$. By Dirac's theorem, H has a Hamiltonian cycle, and so H has a perfect matching. Consequently, in G, we can colour a, b, c and a set F of $n - 1$ independent edges which are not incident with a, b or c with colour c_{2n-1}. By Lemma 4.3, the graph $G \backslash F$ can be edge-coloured with colours $c_1, c_2, \dots, c_{2n-2}$ and by a counting argument it follows that each colour is missing from exactly one vertex in $V(G) \backslash \{a, b, c\}$. Assigning to each such vertex the colour missing at it, we obtain a total colouring of G using $\Delta + 1$ colours. Thus we conclude that if G is Type 2, then $\overline{G}$ does not contain a K_3.

Next we show that G is critical if and only if $\overline{G}$ does not contain a K_3. Since we have already shown that G is Type 2 if and only if $\overline{G}$ does not contain a K_3, we need only show that if $\overline{G}$ does not contain a K_3, then the removal of any edge from G results in a Type 1 graph.

Suppose that $\overline{G}$ does not contain a K_3. If $2 \leq n \leq 4$ then G is graph 4, 10, 30 or 31 in our catalogue, and so is critical. So we assume that $n \geq 5$.

Let $e = v_1v_2$ be an edge in $E(G)$. The graph $\overline{G}$ is regular of degree 2, so there are independent edges $e_1 = v_1w_1$ and $e_2 = v_2w_2$ in $E(\overline{G})$. Create a graph G^* from G by adding a new vertex v^* and joining it to all vertices of G except v_1 and v_2. Then each vertex of G^* has degree $2n - 1$, except for v_1 and v_2 which have degree $2n - 2$. Let F_1 be a 1-factor of G^* containing the edges e and w_1v^*, and let F_2 be a 1-factor of G^* containing the edges e and w_2v^*, and such that $F_1 \cap F_2 = \{e\}$. Recalling that $n \geq 5$, it follows from Lemma 4.5 that G^* has a perfect matching containing e, and w_1v^*, and so the 1-factor F_1 exists; it follows by the same argument applied to the graph $G^*\backslash(F_1\backslash e)$ that F_2 exists. The graph $G^*\backslash(F_1 \cup F_2)$ is regular of order $2n + 2$ and degree $2n - 3$, and so by Lemma 4.4, it is edge-colourable with $2n - 3$ colours.

We give the colour c_1 to the vertices v_1 and w_1 and to the edges of F_1 which are contained in $G\backslash e$. We give the colour c_2 to the vertices v_2 and w_2 and to the edges of F_2 which are contained in $G\backslash e$. We edge-colour $G^*\backslash(F_1 \cup F_2)$ with colours $c_3, \ldots, c_{2n-1}$. We then delete the vertex v^* and colour each vertex x in $V(G)\backslash\{v_1, w_1, v_2, w_2\}$ with the colour of the edge xv^*. All the edges and vertices of $G\backslash e$ are now coloured using a colour from $c_1, \ldots, c_{2n-1}$, so we have shown that $G\backslash e$ is Type 1, as required.   □

Note that Theorem 4.6 does not preclude the existence of nonregular critical graphs of order $2n + 1$ and maximum degree $2n - 2$, although we do not believe that there are any.

We need the following result of Niessen and Volkman [**34**], which improves upon an earlier result of Chetwynd and Hilton [**17**]. Let $r(G)$ denote the number of vertices of G of maximum degree. Call a graph G *overfull* if $|E(G)| > \Delta(G)\lfloor|V(G)|/2\rfloor$.

LEMMA 4.7 (Niessen and Volkman [**34**]). *Let $|V(G)| = 2n + 1$, let $\Delta(G) \geq n + 3r(G) - 3$, and let G be $(r(G) - 2)$-edge-connected. Then G is Class 2 if and only if G is overfull.*

We also need a further lemma proved by Chetwynd and Hilton [**16**] (improving earlier work in [**13**]). This was also proved independently by Niessen and Volkman in [**34**].

LEMMA 4.8 (Chetwynd and Hilton, [**16**]). *Let G be a regular simple graph of even order satisfying*

$$d(G) \geq \tfrac{1}{2}(\sqrt{7} - 1)|V(G)|.$$

Then G is 1-factorizable.

Next we show that if $|V| = 2n + 1$ and $\Delta = 2n - 3$ and $\mathrm{def}(G) = 1$, then it is true, for all except possibly a finite number of small values of n, that G is critical if and only if $\overline{G}$ does not contain a K_3. This gives a result similar to that in Theorem 4.6. The proof of Theorem 4.9 forms a convenient model for the more elaborate proof of our main result, Theorem 4.15.

THEOREM 4.9. *Let G be a graph of odd order $2n+1 \geq 39$, deficiency $\mathrm{def}(G) = 1$ and maximum degree $\Delta = 2n-3$. Then G is critical if and only if $\overline{G}$ does not contain a K_3, i.e., if and only if G is nonconformable.*

PROOF. Since $\mathrm{def}(G) = 1$, G is conformable if and only if $V(G)$ can be coloured by $\Delta + 1 = 2n - 2$ colours in such a way that all, except possibly one, of the colour classes are odd. Since

$$2n + 1 = (2n - 3) \cdot 1 + 1 \cdot 4 = (2n - 4) \cdot 1 + 1 \cdot 2 + 1 \cdot 3$$
$$= (2n - 4) \cdot 1 + 0 \cdot 2 + 1 \cdot 5,$$

this is possible if and only if G contains an independent set of three vertices, i.e., $\overline{G}$ contains a K_3.

We shall show initially that G is Type 2 if and only if $\overline{G}$ does not contain a K_3.

If $\overline{G}$ does not contain a K_3, then G is nonconformable and so is Type 2. Conversely, suppose that $\overline{G}$ contains a K_3. Since we suppose that $2n + 1 \geq 39$, it follows that $n \geq 19$. Let a, b, c be the three vertices of a K_3 in $\overline{G}$ with $d_G(a) = d_G(b) = \Delta = 2n - 3$. Let x be the vertex of G of degree $2n - 4$. Choose y and z so that $yz \in E(\overline{G})$ and $y, z \notin \{a, b, c, x\}$. Note that it is possible that $x = c$. It is easy to see that y and z can be chosen in this way. Let $H = (G \backslash a) \cup bc$. Then H has minimum degree at least $2n - 5$. It follows from Lemma 4.5 that H has a perfect matching containing the edge bc. Consequently, there is a set F_1 in G of $n - 1$ independent edges, all nonincident with a, b and c. We colour a, b, c and the edges of F_1 with the colour c_{2n-2}. Thus if $x = c$, then x has the colour c_{2n-2}. Now let $H' = ((G \backslash F_1) \backslash x) \cup yz$. Then H' also has minimum degree at least $2n - 5$, and so, similarly, there is a set F_2 in $G \backslash F_1$ of $n - 1$ independent edges, all nonincident with x, y and z. We colour y, z and the edges in F_2 with the colour c_{2n-3}. Now consider the graph $G \backslash (F_1 \cup F_2)$. It has five vertices of maximum degree $2n - 4$ and $2n - 4$ vertices of degree $2n - 5$. If $x \neq c$, then x now has degree $2n - 5$, and if $x = c$ then x has degree $2n - 4$. Since $n \geq 19$ it follows that $\Delta\big(G \backslash (F_1 \cup F_2)\big) \geq n + 3 \cdot 5 - 3$, and since

$$\big|E(G) \backslash (F_1 \cup F_2)\big| = \tfrac{1}{2}\big\{5 \cdot (2n - 4) + (2n - 4) \cdot (2n - 5)\big\}$$
$$= n \cdot (2n - 4)$$
$$= \Delta(G \backslash (F_1 \cup F_2)) \cdot \big\lfloor \tfrac{1}{2}|V(G)| \big\rfloor.$$

Since $\delta\big(G \backslash (F_1 \cup F_2)\big) = 2n - 5$ and $\big|V\big(G \backslash (F_1 \cup F_2)\big)\big| = 2n + 1 \geq 39$, $G \backslash (F_1 \cup F_2)$ is well-connected, certainly 3-connected. It follows from Lemma 4.7 that $G \backslash (F_1 \cup F_2)$ can be edge-coloured with $2n - 4$ colours $c_1, c_2, \ldots, c_{2n-4}$. At this point all the edges of G are coloured, and all the vertices a, b, c, y and z are coloured. The colour c_{2n-3} is missing at x. Each of the vertices of $V(G) \backslash \{a, b, c, y, z\}$ has one colour from $\{c_1, \ldots, c_{2n-4}\}$ missing and by counting it is easy to see that each of these colours is missing at exactly one vertex. Therefore each vertex in $V(G) \backslash \{a, b, c, y, z\}$ can be coloured with the colour from $\{c_1, \ldots, c_{2n-4}\}$ missing at it; in particular, if $x \neq c$ then x is coloured with one these colours. This gives the required total colouring of G using $2n - 2$ colours. Thus it follows that if G is Type 2, then $\overline{G}$ cannot contain a K_3.

Next we show that G is critical if and only if $\overline{G}$ does not contain a K_3.

Since we have already shown that G is Type 2 if and only if $\overline{G}$ does not contain a K_3, we need only show that if $\overline{G}$ does not contain a K_3, the removal of any edge from G results in a Type 1 graph.

Suppose that $\overline{G}$ does not contain a K_3. Let e be an arbitrary edge of G. Since $n \geq 19$, it follows that $2n - 5 \geq (\sqrt{7} - 1)(2n + 2)/2$.

Let $e = v_1 v_2$. The graph $\overline{G}$ has one vertex of degree 4, and the remainder have degree 3. Let $v_1 w_1$ and $v_2 w_2$ be independent edges of $\overline{G}$ and let x be the vertex of G of degree $2n - 4$. It is possible to choose w_1 and w_2 so that x, w_1, w_2 are distinct. While we can choose the labels v_1 and v_2 so that $x \neq v_2$, it remains possible that $x = v_1$.

Firstly, suppose that $x \neq v_1$. Let xy be an edge of $\overline{G}$. Since $\overline{G}$ does not contain a K_3, it is easy to see that y can be chosen so that v_1, v_2, w_1, w_2 are different from y. Create a graph G^* from $G \backslash e$ by adding a new vertex v^* and joining it to all vertices of G except w_1, w_2 and y, and by inserting an edge f joining x and y, an edge e_1 joining v_1 and w_1, and an edge e_2 joining v_2 and w_2. The graph G^* is regular of degree $2n - 2$ and order $2n + 2$. Let F_1 be a 1-factor of G^* containing the edges e_1 and xv^*, and not containing any of the edges f, e_2, $v_2 v^*$ or $v_1 v^*$. It follows from Lemma 4.5 that F_1 exists. Let F_2 be a 1-factor of $G^* \backslash F_1$ containing the edges e_2 and $v_1 v^*$, and not containing either of the edges f or $v_2 v^*$. The existence of F_2 follows in the same way by applying Lemma 4.5 to the graph $G^* \backslash F_1$. Finally let F_3 be a 1-factor of $G^* \backslash (F_1 \cup F_2)$ containing the edges f and $v_2 v^*$. This time the existence follows by applying Lemma 4.5 to $G^* \backslash (F_1 \cup F_2)$. The graph $G^* \backslash (F_1 \cup F_2 \cup F_3)$ is regular of degree $2n - 5$. By Lemma 4.8, since $2n - 5 \geq (1/2)(\sqrt{7} - 1)(2n + 2)$, it can be edge-coloured with $c_4, \ldots, c_{2n-2}$.

Using the edge-colouring of $G^* \backslash (F_1 \cup F_2 \cup F_3)$, together with F_1, F_2 and F_3 themselves, we can construct the required total colouring of G. We colour the vertices v_1 and w_1 and all the edges of $F_1 \cap E(G)$ with colour c_1. We colour the vertices v_2 and w_2 and all the edges of $F_2 \cap E(G)$ with colour c_2. We colour the vertices x and y and all the edges of $F_3 \cap E(G)$ with colour c_3. Finally we colour all edges of $(G \backslash e) \backslash (F_1 \cup F_2 \cup F_3)$ with the colour they received in the edge-colouring of $G^* \backslash (F_1 \cup F_2 \cup F_3)$ and we colour each vertex v in $V(G) \backslash \{x, y, v_1, v_2, w_1, w_2\}$ with the colour that the edge vv^* received in the edge-colouring of $G^* \backslash (F_1 \cup F_2 \cup F_3)$.

Secondly suppose that $x = v_1$. We choose vertices y and z such that y, $z \neq w_1$, w_2, v_1, v_2 and yz is an edge of $\overline{G}$. This is possible since $|V(G)| \geq 39$ and $d_{\overline{G}}(u) = 3$ ($u \neq x$), $d_{\overline{G}}(x) = 4$. Construct G^* from $G \backslash e$ as before. This time since x and y are not necessarily nonadjacent in G, G^* may have two edges joining x and y. G^* is again regular of degree $2n - 2$ and order $2n + 2$.

Let F_1 be a 1-factor of G^* containing the edges e_1 and $v_2 v^*$, and not containing any of the edges f, e_2, xv^* or zv^*. Let F_2 be a 1-factor of $G^* \backslash F_1$ containing the edges e_2 and xv^*, and not containing either of the edges f or zv^*. Let F_3 be a 1-factor of $G^* \backslash (F_1 \cup F_2)$ containing the edges f and zv^*. The existence of F_1, F_2 and F_3 follows from Lemma 4.5 as before. Also, as before, $G^* \backslash (F_1 \cup F_2 \cup F_3)$ is regular of degree $2n - 5$, and can be edge-coloured with $c_4, \ldots, c_{2n-2}$.

The total colouring of G with $2n - 2$ colours is obtained as follows: The vertices $v_1 = x$ and w_1 and all the edges of $F_1 \cap E(G)$ are coloured c_1. The vertices v_2 and w_2 and all the edges of $F_2 \cap E(G)$ are coloured c_2. The vertices y and z and all the edges of $F_3 \cap E(G)$ are coloured c_3. The remaining edges and vertices are coloured by the method described for the earlier case. $\square$

The following theorem is due to Chetwynd, Hilton and Zhao Cheng [18] and was proved before the Conformability Conjecture was formulated.

THEOREM 4.10 (Chetwynd, Hilton, Zhao [18]). *Let G be a regular graph of odd order $2n + 1$, satisfying*

$$d = d(G) \geq (1/2)\sqrt{7}(2n + 1).$$

Then G is Type 1 if and only if, for some $s \geq 0$, $\overline{G}$ contains s vertex disjoint odd order subgraphs $K_{i_1}, K_{i_2}, \ldots, K_{i_s}$ with $i_1 + \cdots + i_s = 2n - d + s$, where $i_1 \geq \cdots \geq i_s \geq 3$.

Note that if $|V(G)| = 2n + 1$ and G is conformable and regular, then $d(G) + 1$ colours must each occur on an odd number of vertices. If s colours occur on more than one vertex, say on $i_1, i_2, \ldots, i_s$ vertices, respectively, and the remaining $d(G) + 1 - s$ colours each occur on one vertex, then

$$i_1 + \cdots + i_s + \big(d(G) + 1 - s\big) = 2n + 1,$$

so

$$i_1 + \cdots + i_s = 2n - d(G) + s.$$

We note the following description of conformability if $|V(G)|$ is odd:

LEMMA 4.11. *Let G be a graph of odd order $2n + 1$. Then G is conformable if and only if for some integer t with $0 \leq t \leq \mathrm{def}(G)$, some even integers $j_1 \geq j_2 \geq \ldots j_t \geq 2$, some integer $s \geq 0$, and some odd integers $i_1 \geq \ldots i_s \geq 3$ such that*

$$j_1 + \cdots + j_t + i_1 + \cdots + i_s + \Delta(G) + 1 - s - t = 2n + 1,$$

the graph $\overline{G}$ contains vertex disjoint complete subgraphs $K_{j_1}, \ldots, K_{j_t}, K_{i_1}, \ldots, K_{i_s}$.

PROOF. If G is conformable, then $|V(G)|$ can be coloured with $\Delta(G) + 1$ colours in such a way that the number of colour classes of even order is at most $\mathrm{def}(G)$. We may assume that each colour class of even order contains at least two vertices (for if some colour were not actually used, we could divide some other colour class into two).

If $\overline{G}$ contains vertex disjoint complete subgraphs as described in the statement of the theorem, then the vertex-colouring of G having the vertex-sets of each of the $s + t$ complete graphs in $\overline{G}$ as colour classes and all the remaining vertices in singleton colour classes, is conformable. Lemma 4.11 follows. $\square$

In view of Lemma 4.11, it is worth noting that Theorem 4.10 is really stating that G is Type 1 if and only if G is conformable. Since G is a regular graph of degree larger than half its order, G contains no nonconformable subgraph having the same maximum degree (this result is easily shown). Theorem 4.10, therefore, provides additional evidence for the Conformability Conjecture.

Before establishing the main results of this section, Theorem 4.15 and Corollary 4.16, we need a few lemmas.

LEMMA 4.12. *Let G be a graph having odd order $2n + 1$ and maximum degree Δ with $2n - 1 \geq \Delta \geq 3(2n)/4 + \mathrm{def}(G)/4$. Let $\varphi: V \to \{1, 2, \ldots, \Delta + 1\}$ be a vertex colouring with colour classes $V_i = \varphi^{-1}(i)$ for $i = 1, 2, \ldots, \Delta + 1$. If every V_i is nonempty, then there exist two singleton colour classes $V_j = \{z_j\}$, $V_k = \{z_k\}$ for which z_j and z_k are nonadjacent.*

PROOF. Let W be the union of all nonsingleton colour classes and let $\overline{W} = V \backslash W$.

If s is the number of nonsingleton odd order colour classes and t is the number of even order colour classes, it follows that $|\overline{W}| = \Delta + 1 - s - t$ and thus that

$$|\overline{W}| - |W| = |\overline{W}| - (2n + 1 - |\overline{W}|) = 2(\Delta + 1 - s - t) - (2n + 1).$$

Counting vertices, and recalling that no colour classes are empty, we see that

$$3s + 2t + (\Delta + 1 - s - t) \leq 2n + 1,$$

so

$$2s + t \leq 2n - \Delta,$$

or crudely, since $s \geq 0$,

$$s + t \leq 2n - \Delta.$$

Therefore

$$|\overline{W}| - |W| \geq 4\Delta - 6n + 1.$$

The number of edges of $\overline{G}$ joining vertices in W to vertices in $\overline{W}$ is at most $\delta(\overline{G})|W| + \operatorname{def}(G)$. Therefore the number of edges of $\overline{G}$ joining two vertices of $\overline{W}$ is at least

$$e = \tfrac{1}{2}\bigl(\delta(\overline{G})|\overline{W}| - \delta(\overline{G})|W| - \operatorname{def}(G)\bigr) = \tfrac{1}{2}\delta(\overline{G})\bigl(|\overline{W}| - |W|\bigr) - \tfrac{1}{2}\operatorname{def}(G).$$

Recall that $2n - 1 \geq \Delta$. Since $\delta(\overline{G}) = 2n - \Delta \geq 1$, the number of edges of $\overline{G}$ joining two vertices of $\overline{W}$ is at least

$$e \geq \tfrac{1}{2}(1)(4\Delta - 6n + 1) - \tfrac{1}{2}\operatorname{def}(G).$$

Recall that $\Delta \geq 3(2n)/4 + \operatorname{def}(G)/4$. It follows that

$$e \geq \tfrac{1}{2}(1)\bigl(\operatorname{def}(G) + 1\bigr) - \tfrac{1}{2}\operatorname{def}(G) > 0.$$

Thus there exists at least one edge in $\overline{G}$ joining two vertices in $\overline{W}$, proving the result. $\square$

LEMMA 4.13. *Let G be a graph having odd order $2n + 1$, deficiency $\operatorname{def}(G)$, and maximum degree Δ where $2n - 1 \geq \Delta \geq 3(2n)/4 + \operatorname{def}(G)/4$. Assume that $\operatorname{def}(G) \leq 2n - \Delta$ and G has a conformable vertex-colouring with s nonsingleton odd colour classes. Then G has a conformable vertex-colouring with $\operatorname{def}(G)$ even colour classes, no empty colour classes, and at most s nonsingleton odd colour classes.*

PROOF. Let

$$\varphi \colon V(G) \to \{1, 2, \dots, \Delta + 1\}$$

be a conformable vertex-colouring of G.

If φ has an empty colour class then it must also contain a colour class having $\ell \geq 3$ vertices, for otherwise G would contain at most $2\bigl(\operatorname{def}(G) - 1\bigr) + \Delta - \bigl(\operatorname{def}(G) - 1\bigr) = \operatorname{def}(G) + \Delta - 1 \leq 2n - \Delta + \Delta - 1 = 2n - 1$ vertices, a contradiction. We can then replace the empty colour class and the colour class containing ℓ vertices by a colour class containing two vertices and a colour class containing $\ell - 2$ vertices, without changing the number of even colour classes and without increasing s.

Suppose that φ has t even colour classes. Since $|V(G)|$ is odd, the number of odd colour classes must be odd, so $\Delta + 1 - t$ is odd, so $\Delta \equiv t \bmod 2$. By Lemma 3.5, $\Delta \equiv \operatorname{def}(G) \pmod 2$, so $t \equiv \operatorname{def}(G) \pmod 2$.

If $t \leq \operatorname{def}(G) - 2$, then the conditions of Lemma 4.12 are satisfied by φ. Thus there is a pair of nonadjacent vertices, each of which occurs in a singleton colour class. Putting both these vertices in a single colour class increases the value of t by exactly two (one of the new even colour classes is the empty colour class) and does not change s.

Repeating these two manoeuvres a finite number of times produces the required conformable vertex-colouring. $\qquad\square$

Lemma 4.14 shows that if G is conformable and $|V(G)|$ is odd, then it is possible to choose a conformable vertex-colouring so that for each vertex colour class of the "wrong" parity, there is an associated vertex of degree less than $\Delta(G)$ that is not in that colour class. The idea is that this vertex will be missing the colour assigned to the corresponding colour class in some $(\Delta(G) + 1)$-total colouring extending the conformable vertex colouring.

LEMMA 4.14. *Let G be a graph having odd order $2n + 1$, deficiency $\operatorname{def}(G)$ and maximum degree Δ satisfying $2n - 1 \geq \Delta \geq 3(2n)/4 + \operatorname{def}(G)/4$. If G has a conformable vertex-colouring $\varphi \colon V \to \{1, 2, \ldots, \Delta + 1\}$ with s nonsingleton odd colour classes, t even colour classes and no empty colour classes, then G has a conformable vertex-colouring $\varphi' \to V \to \{1, 2, \ldots, \Delta + 1\}$ with at most $s + 1$ nonsingleton odd colour classes, t even colour classes, $V_1', V_2', \ldots, V_t'$ say, no empty colour classes and there is a multiset of vertices $\{y_1, y_2, \ldots, y_t\}$ such that $y_i \notin V_i'$ for $i = 1, 2, \ldots, t$ where a vertex y appears in the multiset $\{y_1, y_2, \ldots, y_t\}$ at most $\Delta - d_G(y)$ times.*

PROOF. Let $\varphi \colon V \to \{1, 2, \ldots, \Delta + 1\}$ be a conformable vertex colouring with t even colour classes, $V_i = \varphi^{-1}(i)$ for $i = 1, 2, \ldots, t$, and s nonsingleton odd colour classes, $V_i = \varphi^{-1}(i)$ for $i = t + 1, t + 2, \ldots, t + s$. Let $\{x_1, x_2, \ldots, x_{\operatorname{def}(G)}\}$ be a multiset containing all vertices $x \in V(G)$ having $d_G(x) < \Delta$, such that x occurs exactly $\Delta - d_G(x)$ times in the multiset.

The lemma is clearly true when $t = 0$. Now suppose that $t \geq 1$. We begin the proof by showing that there exists a submultiset $\{y_1, y_2, \ldots, y_t\}$ of the multiset $\{x_1, x_2, \ldots, x_{\operatorname{def}(G)}\}$ for which $y_j \notin V_j$ for $1 \leq j \leq t$, unless $\{x \colon d_G(x) < \Delta\} \subseteq V_{j_0}$ for some $1 \leq j_0 \leq t$.

Initially, assume that $\{x \colon d_G(x) < \Delta\} \not\subseteq V_j$ for any $1 \leq j \leq t$. Construct a bipartite graph B having $\{x_1, x_2, \ldots, x_{\operatorname{def}(G)}\} \cup \{V_1, V_2, \ldots, V_t\}$ as its vertex-set, where x_ℓ is joined to V_j if $x_\ell \notin V_j$. Clearly $d_B(x_\ell) = t$ if x_ℓ is not in any of the even colour classes and $d_B(x_\ell) = t - 1$, otherwise. Thus

$$\left| \bigcup_{\ell \in L} N_B(x_\ell) \right| \geq |L|$$

whenever $|L| < t$. If $|L| = t$, then this inequality is also true provided $\{x_\ell \colon \ell \in L \text{ and } d_G(x_\ell) < \Delta\} \not\subseteq V_j$ for any $1 \leq j \leq t$. The assumption implies that there is a choice L_0 of the index set L having $|L_0| = t$ for which $\{x_\ell \colon \ell \in L_0 \text{ and } d_G(x_\ell) < \Delta\} \not\subseteq V_j$ for any $1 \leq j \leq t$. Let $\{y_1, y_2, \ldots, y_t\}$ be the multiset consisting of the vertices x_ℓ for $\ell \in L_0$. By Hall's Theorem, the vertex induced bipartite subgraph B' having vertex-set $\{y_1, y_2, \ldots, y_t\} \cup \{V_1, V_2, \ldots, V_t\}$ has a perfect matching and so we can order the vertices $y_1, y_2, \ldots, y_t$ so that $y_j \notin V_j$ for $1 \leq j \leq t$.

Next, we show that if $\{x \colon d_G(x) < \Delta\} \subseteq V_{j_0}$ for some $1 \leq j_0 \leq t$, then it is possible to obtain from φ a conformable vertex-colouring φ' that contains t even colour classes and either s or $s + 1$ nonsingleton odd colour classes, and such that $\{x \colon d_G(x) < \Delta\}$ is not a subset of any V_j.

Since φ has no empty colour classes, Lemma 4.12 guarantees the existence of a pair of singleton colour classes, $V_{t+s+1} = \{z_1\}$ and $V_{t+s+2} = \{z_2\}$ say, for which $z_1 z_2 \notin E(G)$.

If $|V_{j_0}| = 2$, say $V_{j_0} = \{u_1, u_2\}$, then we define $V'_{j_0} = \{z_1, z_2\}$, $V'_{t+s+1} = \{u_1\}$ and $V'_{t+s+2} = \{u_2\}$. If $|V_{j_0}| \geq 4$, then we remove one vertex, u say, from V_{j_0}, and let $V'_{t+s+1} = V_{j_0} \backslash \{u\}$, $V'_{t+s+2} = \{u\}$ and $V'_{j_0} = \{z_1, z_2\}$. We take care that u is a vertex in V_{j_0} such that $d_G(u) < \Delta$.

The colouring φ' corresponding to the colour classes V'_i (where $V'_i = V_i$ if $i \neq j_0$, $t + s + 1$ and $t + s + 2$) has the property that there does not exist a $1 \leq j \leq t$ for which $\{x \colon d_G(x) < \Delta\} \subseteq V'_j$. Furthermore φ' has t even order colour classes and either s or $s + 1$ nonsingleton odd colour classes and no empty colour classes. Thus applying the earlier argument to the colouring φ' guarantees the existence of the multiset $\{y_1, y_2, \ldots, y_t\}$. $\qquad\square$

We are now ready to prove the key result of this section, Theorem 4.15.

THEOREM 4.15. *Let G be a graph of odd order $2n + 1$, deficiency $\mathrm{def}(G)$ and maximum degree Δ. Suppose that G has a conformable vertex-colouring with s nonsingleton odd order colour classes. If*

$$\Delta \geq (\sqrt{7} - 1)n + \mathrm{def}(G) + s + (\sqrt{7} - 1),$$

and

$$\mathrm{def}(G) \leq 2n - \Delta,$$

then G is Type 1.

Note that $(\sqrt{7} - 1)n \sim 0.823\big(|V(G)| - 1\big)$.

PROOF. Since K_{2n+1} is Type 1, so is any graph having odd order $2n + 1$ and maximum degree $\Delta = 2n$. Thus we may assume that $\Delta \leq 2n - 1$. Since $\Delta \leq 2n - 1$ and $\Delta \geq (\sqrt{7} - 1)n + (\sqrt{7} - 1)$, we may assume throughout this proof that $n \geq 8$.

Since

$$(\sqrt{7} - 1)n + \mathrm{def}(G) + s + (\sqrt{7} - 1) \geq \tfrac{3}{4}(2n) + \tfrac{1}{4}\mathrm{def}(G)$$

it follows from Lemma 4.13 that G has a conformable vertex-colouring with $\mathrm{def}(G)$ even colour classes, no empty colour classes, and at most s nonsingleton odd colour classes. By Lemma 4.14 G has a conformable vertex-colouring $\varphi \colon V \to \{1, 2, \ldots, \Delta + 1\}$ with at most $s + 1$ nonsingleton odd colour classes, $\mathrm{def}(G)$ even colour classes $V_1, \ldots, V_{\mathrm{def}(G)}$, and no empty colour classes, for which there exists a multiset of vertices $\{y_1, \ldots, y_{\mathrm{def}(G)}\}$ such that $y_i \notin V_i$ for $i = 1, 2, \ldots, \mathrm{def}(G)$, where a vertex y appears in the multiset $\{y_1, \ldots, y_{\mathrm{def}(G)}\}$ at most $\Delta - d_G(y)$ times.

Let $W_1, \ldots, W_{s'}$ be the odd nonsingleton colour classes of φ. Then $s \geq s' - 1$ so it follows from the hypothesis that

$$\Delta \geq (\sqrt{7} - 1)n + \mathrm{def}(G) + s' + (\sqrt{7} - 2).$$

Now we proceed to find a total colouring of G with $\Delta + 1$ colours.

From G we create a multigraph H as follows: For $1 \leq \ell \leq s'$, let z_ℓ be a vertex of W_ℓ and let M_ℓ be a 1-factor in the complete graph induced by the set $W_\ell \backslash z_\ell$ in the complement of G. For $1 \leq \ell \leq \mathrm{def}(G)$, let N_ℓ be a 1-factor in the complete graph induced by V_ℓ. Let G' denote the graph G with the extra edges of $M_1 \cup \cdots \cup M_{s'} \cup N_1 \cup \cdots \cup N_{\text{def}(G)}$ adjoined. Add a new vertex v^* to G', joining v^* to each vertex $v \in V(G)$ for which $d_{G'}(v) < \Delta + 1$ by $(\Delta + 1) - d_{G'}(v)$ edges. The multigraph formed in this manner is called H.

The multigraph H is regular of order $2n + 2$ and degree $\Delta + 1$; the degree of v^* is calculated using the equality $j_1 + \cdots + j_{\text{def}(G)} + i_1 + \cdots + i_{s'} + \Delta + 1 - s' - \text{def}(G) = 2n + 1$ (see Lemma 4.11).

From H, we pick out edge-disjoint 1-factors $F_1, F_2, \ldots, F_{s'+\text{def}(G)}$ having the following properties: For $1 \leq \ell \leq s'$, F_ℓ is a 1-factor of $H\backslash(F_1 \cup F_2 \cup \cdots \cup F_{\ell-1})$ containing the partial matching M_ℓ and the edge $v^* z_\ell$. Let $M_\ell^* = M_\ell \cup \{v^* z_\ell\}$. Let G_ℓ denote the simple graph underlying the multigraph $H\backslash(F_1 \cup \cdots \cup F_{\ell-1} \cup M_{\ell+1} \cup \cdots \cup M_{s'} \cup N_1 \cup \cdots \cup N_{\text{def}(G)})$. To apply Lemma 4.5 to show the existence of F_ℓ, we need only show that $\delta(G_\ell) \geq |V(G_\ell)|/2 + |M_\ell^*|$ since there are no forbidden edges. Notice that

$$\delta(G_\ell) \geq \Delta + 1 - (\ell - 1) - 1 - \text{def}(G) = \Delta - \ell - \text{def}(G) + 1,$$

and that $|V(G_\ell)| = 2n + 2$ and $|M_\ell^*| = (|W_\ell| - 1)/2 + 1$. Therefore, F_ℓ exists, provided

$$\Delta - \ell - \text{def}(G) + 1 \geq n + 1 + \tfrac{1}{2}(|W_\ell| - 1) + 1$$

or

$$2(\Delta - n - \ell - \text{def}(G)) - 1 \geq |W_\ell|.$$

Since

$$\Delta - n - \ell - \text{def}(G) \geq ((\sqrt{7} - 1)n + s' + \text{def}(G) + (\sqrt{7} - 2)) - n - \ell - \text{def}(G)$$
$$\geq (\sqrt{7} - 2)n + (\sqrt{7} - 2) \geq 0.6n,$$

and, using a very crude bound,

$$|W_\ell| \leq (2n + 1) - (\Delta + 1) \leq 2n - ((\sqrt{7} - 1)n + s' + \text{def}(G) + (\sqrt{7} - 2))$$
$$\leq (3 - \sqrt{7})n \leq 0.4n,$$

the conditions of Lemma 4.5 are satisfied because we assume that $n \geq 8$. Thus the required matching F_ℓ exists.

For $s' + 1 \leq \ell \leq s' + \text{def}(G)$, F_ℓ is a 1-factor of $H\backslash(F_1 \cup F_2 \cup \cdots \cup F_{\ell-1})$ containing the partial matching $N_{\ell-s'}$ and an edge joining v^* to $y_{\ell-s'}$. Let $N_{\ell-s'}^* = N_{\ell-s'} \cup \{v^* y_{\ell-s'}\}$. Recall that $y_{\ell-s'} \notin V_{\ell-s'}$ so that $N_{\ell-s}^*$ is a partial matching. Let G_ℓ denote the simple graph underlying $H\backslash(F_1 \cup \cdots \cup F_{\ell-1} \cup N_{\ell-s'+1} \cup \cdots \cup N_{\text{def}(G)})$. Once again, to apply Lemma 4.5 we need only show that $\delta(G_\ell) \geq |V(G_\ell)|/2 + |N_{\ell-s'}^*|$. By the same argument as before, since $|N_{\ell-s'}^*| = |V_\ell|/2 + 1$, it follows that F_ℓ exists provided

$$2(\Delta - n - \ell - \text{def}(G) - 1) \geq |V_\ell|.$$

Using the same argument as that given above, we can deduce that $(\Delta - n - \ell - \text{def}(G)) \geq 0.6n$ while $|V_\ell| \leq 0.4n$, and thus conclude that the matching F_ℓ exists.

Now consider the graph $H' = H\backslash(F_1 \cup \cdots \cup F_{s'+\text{def}(G)})$. The graph H' contains no multiple edges since $F_1 \cup \cdots \cup F_{s'+\text{def}(G)}$ contains all the edges $y_i v^*$ ($1 \leq i \leq \text{def}(G)$). It is regular of degree $\Delta + 1 - s' - \text{def}(G) \geq (\sqrt{7} - 1)n + (\sqrt{7} - 1)$ and so by Lemma 4.8 is Class 1.

We obtain a total colouring of G, with $\Delta+1$ colours, as follows: For $1 \leq \ell \leq s'$, the vertices of W_ℓ and the edges of F_ℓ which are in $G \backslash W_\ell$ are coloured with colour c_ℓ. For $s'+1 \leq \ell \leq \mathrm{def}(G)$, the vertices of $V_{\ell-s'}$ and the edges of F_ℓ which are in $G \backslash (V_{\ell-s'} \cup \{y_{\ell-s'}\})$ are coloured with c_ℓ. We edge-colour H' with colours $c_{s'+\mathrm{def}(G)+1}, \ldots, c_{\Delta+1}$, then colour each edge of G which is also in H' with the colour it received in H'. For each $v \in V(G)$ with $vv^* \in E(H')$, colour v with the colour the edge vv^* receives in H'. This yields a total colouring of G with $\Delta+1$ colours. Thus G is Type 1, as required. $\qquad\square$

The next result (which generalizes Theorem 4.10) is a corollary to Theorem 4.15 and Lemma 4.13. Although the statement is rather nicer than that of Theorem 4.15 itself, the lower bound for Δ in the hypothesis is much larger.

COROLLARY 4.16. *Let G be a graph of odd order $2n+1$, maximum degree*

$$\Delta \geq \tfrac{1}{3}\big[\sqrt{7}(2n+1) + \mathrm{def}(G) + (\sqrt{7}-2)\big]$$

and deficiency

$$\mathrm{def}(G) \leq 2n - \Delta.$$

Then G is Type 1 if and only if G is conformable.

Note: $\sqrt{7}/3 \sim 0.882$

PROOF. Suppose that G is Type 1; then Proposition 2.1 shows that G is conformable.

Conversely, suppose that G is conformable. If $\Delta = 2n$ then $\mathrm{def}(G) = 0$, so G is regular, so $G = K_{2n+1}$. Then G is Type 1. If $2n-1 \geq \Delta$, then by Lemma 4.13, since

$$\tfrac{1}{3}\big[\sqrt{7}(2n+1) + \mathrm{def}(G) + (\sqrt{7}-2)\big] \geq \tfrac{3}{4}(2n) + \tfrac{1}{4}\mathrm{def}(G)$$

and

$$\mathrm{def}(G) \leq 2n - \Delta,$$

it follows that G has a conformable vertex-colouring φ with $\mathrm{def}(G)$ even colour classes and no empty colour classes. Let φ have s odd nonsingleton colour classes. Then

$$2\,\mathrm{def}(G) + 3s + \big(\Delta + 1 - \mathrm{def}(G) - s\big) \leq 2n+1,$$

so that

$$\mathrm{def}(G) + 2s + \Delta + 1 \leq 2n+1,$$

and thus

$$s \leq \tfrac{1}{2}\big(2n - \Delta - \mathrm{def}(G)\big).$$

Since

$$\Delta \geq \tfrac{1}{3}\big[\sqrt{7}(2n+1) + \mathrm{def}(G) + (\sqrt{7}-2)\big],$$

it follows that

$$\tfrac{3}{2}\Delta \geq \sqrt{7}n + \tfrac{1}{2}\mathrm{def}(G) + \sqrt{7} - 1,$$

so that

$$\Delta \geq (\sqrt{7} - 1)n + \operatorname{def}(G) + (\sqrt{7} - 1) + \tfrac{1}{2}\left(2n - \Delta - \operatorname{def}(G)\right)$$
$$\geq (\sqrt{7} - 1)n + \operatorname{def}(G) + (\sqrt{7} - 1) + s,$$

so, by Theorem 4.15, G is Type 1.   □

In view of Lemma 3.8 and Theorem 3.9 about graphs of even order, it is no surprise that Corollary 4.16 simply requires that the graph G itself must be conformable if it is to be Type 1, and does not also involve a requirement that the subgraphs of G of maximum degree $\Delta(G)$ should also be conformable. The reason (which can be deduced from Corollary 4.16) is that if G is conformable and has sufficiently large maximum degree and sufficiently small deficiency, then G cannot have nonconformable subgraphs of the same maximum degree. The next result establishes this fact for odd order graphs with a smaller lower bound for the maximum degree than that given in Corollary 4.16.

PROPOSITION 4.17. *Let G be a graph having odd order $2n+1$, maximum degree $\Delta \geq 3(2n)/4 + \operatorname{def}(G)/4$ and deficiency $\operatorname{def}(G) \leq 2n - \Delta$. If G is conformable, then G does not have any nonconformable subgraphs of maximum degree Δ.*

PROOF. If $\Delta = 2n$ then $G = K_{2n+1}$, G is conformable and is Type 1, so G has no nonconformable subgraphs of maximum degree Δ. If $2n - 1 \geq \Delta$ then, by Lemma 4.13, G has a conformable vertex-colouring

$$\varphi \colon V \to \{1, 2, \ldots, \Delta + 1\}$$

with $\operatorname{def}(G)$ even colour classes and no empty colour classes.

It is sufficient to show that G has no proper induced subgraphs H of maximum degree Δ which are nonconformable. This follows from Lemma 3.8 if $|V(H)| \leq |V(G)| - 2$. Now let $w \in V(G)$, let $H = G\backslash\{w\}$, and let φ_H be the vertex colouring of H induced by φ. By $(*)$ in the proof of Lemma 3.8, $\operatorname{def}(H) \geq \Delta - \operatorname{def}(G)$. If w is in one of the odd colour classes of φ, then φ_H has $\Delta + 1 - \operatorname{def}(G) - 1 = \Delta - \operatorname{def}(G) \leq \operatorname{def}(H)$ odd colour classes, so H is conformable. If w is in one of the even colour classes of φ, then, since φ has no empty colour classes, it follows from Lemma 4.12 that there is a pair of nonadjacent vertices in G that occur in singleton colour classes. Let φ' be the conformable vertex colouring of G obtained from φ by splitting the even colour class containing w into two odd colour classes, and putting the nonadjacent pair of vertices into a single even colour class. Since w is now in an odd colour class of φ' it follows as above that the vertex-colouring φ'_H of H induced by φ' is conformable.   □

It is worth making a comment at this point about the structure of the critical graphs having odd order, large maximum degree and small deficiency.

Let a *barely nonconformable* graph be a nonconformable graph G for which there is an integer t, $0 \leq t \leq \operatorname{def}(G)$, even integers $j_1, \ldots, j_t$ greater than zero, and odd integers $i_1, \ldots, i_s$ greater than one, such that

$$j_1 + \cdots + j_t + i_1 + \cdots + i_s + \Delta(G) + 1 - s - t = 2n - 1,$$

and $\overline{G}$ contains disjoint complete subgraphs $K_{j_1}, \ldots, K_{j_t}, K_{i_1}, \ldots, K_{i_s}$. (Note that if the equation above held with $2n - 1$ replaced by $2n + 1$ then, by Lemma 4.11, G would be conformable).

Let a *nearly conformable* graph be a nonconformable graph G for which there is an integer t, $0 \leq t \leq \mathrm{def}(G) + 2$, even integers $j_1, \ldots, j_t$ greater than zero, and odd integers $i_1, \ldots, i_s$ greater than one, such that

$$j_1 + \cdots + j_t + i_1 + \cdots + i_s + \Delta(G) + 1 - s - t = 2n - 1,$$

and $\overline{G}$ contains vertex disjoint complete subgraphs $K_{j_1}, \ldots, K_{j_t}, K_{i_1}, \ldots, K_{i_s}$.

Clearly a barely nonconformable graph is nearly conformable. The properties of being barely nonconformable or nearly conformable are, as the names are meant to suggest, closely associated with the property of being critical.

THEOREM 4.18. *Let G be a graph of odd order $2n + 1$ and maximum degree*

$$\Delta \geq \tfrac{1}{3}\left[\sqrt{7}(2n + 1) + \mathrm{def}(G) + (\sqrt{7} - 2)\right] + 1.$$

(a) *If G is barely nonconformable, then G is critical.*
(b) *If G is critical, then G is nearly conformable.*

PROOF. (a) Suppose that G is barely nonconformable. Then G is nonconformable. By the Hajnal-Szemerédi Theorem [**22**], G has a vertex colouring with $\Delta + 1$ colours that has $2n - \Delta$ doubleton colour classes and $2\Delta - 2n + 1$ singleton colour classes. Thus, if $\mathrm{def}(G) \geq 2n - \Delta$, G is conformable. Since G is in fact nonconformable, it follows that $\mathrm{def}(G) \leq 2n - \Delta - 1$. Since $\mathrm{def}(G) \equiv \Delta(G)$ (mod 2) by Lemma 3.5, we have that $\mathrm{def}(G) \leq 2n - \Delta - 2$. By Proposition 2.1, G is Type 2. Since G is barely nonconformable, there exists a set of vertex disjoint complete graphs in $\overline{G}$, $\mathcal{K} = \{K_{j_1}, \ldots, K_{j_t}, K_{i_1}, \ldots, K_{i_s}\}$, with parameters $s, t, j_1, \ldots, j_t, i_1, \ldots, i_s$ where $j_\ell \geq 2$ and is even and $i_\ell \geq 3$ and is odd, such that

$$\sum_{\ell=1}^{t} j_\ell + \sum_{\ell=1}^{s} i_\ell + (\Delta + 1) - s - t = 2n - 1$$

and $0 \leq t \leq \mathrm{def}(G)$. Consider the new set where we take two independent edges in $\overline{G}$ which are not incident with a vertex in one of the complete graphs in $\mathcal{K}$ and add these edges to $\mathcal{K}$ to form $\mathcal{K}'$; the existence of these edges follows from the arguments used in Lemmas 4.12 and 4.13. Briefly, we can colour all but two of the vertices of G with $\Delta + 1$ colours so that j_k $(1 \leq k \leq t)$ vertices are coloured the same and i_k $(1 \leq k \leq s)$ vertices are coloured the same, the remaining nonempty colour classes being singletons. We can then assume that there are no empty colour classes. Then we use the argument of Lemma 4.12 to show that the two independent edges of $\overline{G}$ exist. To do this we need the fact that $\Delta \geq 3(2n)/4 + \mathrm{def}(G)/4 + 1$. If e is an arbitrary edge of $E(G)$, then $\mathrm{def}(G \backslash e) = \mathrm{def}(G) + 2$. Therefore the new set of vertex disjoint complete graphs $\mathcal{K}'$ has parameters $s', t', j_1', \ldots, j_{t'}', i_1', \ldots, i_{s'}'$ such that $s' = s$, $t' = t + 2 \leq \mathrm{def}(G \backslash e)$, $j_\ell' = j_\ell$ for $1 \leq \ell \leq t$, $j_{t+1}' = j_{t+2}' = 2$, $i_\ell' = i_\ell$ for $1 \leq \ell \leq s$, and

$$\sum_{\ell=1}^{t'} j_\ell' + \sum_{\ell=1}^{s'} i_\ell' + (\Delta + 1) - s' - t' = 2n + 1.$$

Since $\mathrm{def}(G \backslash e) \leq \mathrm{def}(G) + 2 \leq 2n - \Delta$, it follows that $\Delta(G \backslash e) = \Delta(G)$, and then it follows from Lemma 4.11 and Corollary 4.16 that $G \backslash e$ is Type 1. Thus G is critical.

(b) Suppose that G is critical. Then, by Corollary 4.16, G is nonconformable. As in (a), above, $\mathrm{def}(G) \leq 2n - \Delta - 2$. Therefore G has at least $2n + 1 - \mathrm{def}(G) \geq \Delta + 3$ vertices of maximum degree. Let e be an edge of G. Then $\Delta(G \backslash e) = \Delta(G)$

and so, since G is critical, $G \backslash e$ is Type 1. Consequently, by Proposition 2.1 $G \backslash e$ is conformable, so by Lemma 4.11, $\overline{G \backslash e}$ contains vertex disjoint complete graphs

$$K_{j_1}, \ldots, K_{j_t}, K_{i_1}, \ldots, K_{i_s},$$

where $0 \le t \le \mathrm{def}(G \backslash e) = \mathrm{def}(G) + 2$, $j_1, \ldots, j_t$ are even integers greater than zero, $i_1, \ldots, i_s$ are odd integers greater than one, and

$$j_1 + \cdots + j_t + i_1 + \cdots + i_s + \Delta + 1 - s - t = 2n + 1.$$

If $G \backslash e$ contains a complete graph of order greater than two, then delete two vertices of such a complete graph. In the case when the complete graph has even order, then, supposing for example that $j_t \neq 2$, we define $j_t' = j_t - 2$, $j_\ell' = j_\ell$ for $1 \le \ell < t$, and $i_\ell' = i_\ell$ for $1 \le \ell \le s$. Then we have

$$j_1' + \cdots + j_t' + i_1' + \cdots + i_s' + \Delta + 1 - s - t = 2n - 1.$$

If $s > 0$ we can argue similarly with an odd order complete graph. If $\overline{G \backslash e}$ contains no complete graphs of order greater than two, then, since $2t + \Delta + 1 - t = 2n + 1$, we have $2n - \Delta = t \le \mathrm{def}(G) + 2 \le 2n - \Delta$, so it follows that $\Delta \le 2n - 2$. Therefore $t \ge 2$. In this case, delete a vertex of K_{j_t} and a vertex of $K_{j_{t-1}}$. Put $t' = t - 2$, $j_l' = j_l$ $(1 \le l \le t')$ and $s = 0$. Then

$$j_1' + \cdots + j_{t'}' + i_1 + \cdots + i_s + \Delta + 1 - s - t' = j_1' + \cdots + j_{t'}' + \Delta + 1 - t'$$
$$= 2n - 1.$$

In every case, it follows that G is nearly conformable. $\square$

It is worth commenting that there exist critical graphs which are not barely nonconformable. An example of such a graph is the graph G with $V(G) = \{x_1, \ldots, x_{11}\}$ and where the pairs of vertices $x_i x_{i+1}$ ($1 \le i \le 11$ with addition modulo 11) and the pairs of vertices $x_i x_{i+3}$ ($1 \le i \le 11$ with addition modulo 11) are the only nonedges of the graph G.

5. The Bipartite Case

If n is even the graph $K_{n,n}$ is conformable but Type 2. It is because of this that the bound $\Delta(G) \ge \left(|V(G)| + 1 \right) / 2$ in the Conformability Conjecture cannot be lowered. However the reason why $K_{n,n}$ is Type 2 is that there is no way of colouring the vertices of $K_{n,n}$ that is not fairly obviously unsuitable for extension to a total colouring with $\Delta + 1$ colours. Thus the spirit of the Conformability Conjecture remains valid, even if the details are no longer correct.

We call the necessary condition that vertex-colourings of bipartite graphs must satisfy, if they are to be extendible to a Type 1 total colouring, biconformability. We state a conjecture, called the Biconformability Conjecture, proposing a necessary and sufficient condition for a bipartite graph of high maximum degree to be Type 1. It follows from results in [19] that almost all unlabelled bipartite graphs satisfy the condition $\Delta(G) \ge |V(G)|/4$. Therefore the Biconformability Conjecture, if true, would classify almost all bipartite graphs into Type 1 or Type 2 graphs. This would provide an interesting contrast to Sánchez-Arroyo's result, which showed that the problem of determining whether a bipartite graph is Type 1 is NP-hard.

Now let us define our terms. If G is a bipartite graph with bipartition (A, B), let $\tilde{G}$ denote the *bipartite complement* of G, that is to say the bipartite graph with bipartition (A, B) for which ab is an edge of $\tilde{G}$ if and only if ab is not an edge of G.

Call a bipartite graph G *equibipartite* if it has bipartition (A, B) of the vertex-set such that $|A| = |B|$, and each edge joins a vertex of A to a vertex of B.

Given an equibipartite graph G and given a vertex-colouring which assigns the colours $c_1, c_2, \ldots, c_{\Delta(G)+1}$, let A_i be the set of vertices of A coloured c_i and B_i the set of vertices of B coloured c_i. Let $a_i = |A_i|$ and $b_i = |B_i|$. If W is a subset of $V(G)$, let $V_{<\Delta}(W)$ denote the set of vertices in W which have degree less than Δ in the graph G. Call G *biconformable* if G has a vertex-colouring such that for $1 \leq i \leq \Delta(G) + 1$,

$$\left| V_{<\Delta}(A \backslash A_i) \right| \geq b_i - a_i, \quad \left| V_{<\Delta}(B \backslash B_i) \right| \geq a_i - b_i,$$

and

$$\mathrm{def}(G) \geq \sum_{i=1}^{\Delta(G)+1} |a_i - b_i|.$$

Note that this definition of biconformability differs from that proposed by Chetwynd and Hilton in [15] by the inclusion of the inequalities for $\left| V_{<\Delta}(A \backslash A_i) \right|$ and $\left| V_{<\Delta}(B \backslash B_i) \right|$. A modification to the original definition of biconformability was suggested by J. Wojciechowski. A further modification was then suggested by L. Andersen to give the current definition. The proof in [15], that if G is biconformable then G is also conformable, is unaffected by the modification. The fact that the bipartite case is less well understood than the general even and odd order cases may be partly due to the fact that the current definition of biconformability has only recently been arrived at.

LEMMA 5.1. *Let G be an equibipartite graph which is not biconformable. Then G is Type 2.*

PROOF. Suppose that G is Type 1. Consider a total colouring of G with colours $c_1, c_2, \ldots, c_{\Delta(G)+1}$. Recall that if v is a vertex of maximum degree, then each colour c_i is used to colour v or an edge incident with v. Suppose that $a_i \geq b_i$ for some i. Then $a_i - b_i$ vertices of B must have colour c_i missing at them. These vertices are in $B \backslash B_i$, and have degree less than Δ. Consequently, $\left| V_{<\Delta}(B \backslash B_i) \right| \geq a_i - b_i$. Similarly $\left| V_{<\Delta}(A \backslash A_i) \right| \geq b_i - a_i$ for each i. Furthermore, for each i, $|a_i - b_i|$ vertices have colour c_i missing at them, so $\mathrm{def}(G) \geq \sum_{i=1}^{\Delta(G)+1} |a_i - b_i|$. Thus G is biconformable and the lemma follows. □

This suggests the following conjecture, originally proposed by Chetwynd and Hilton in [15]. The value 3/14 may have to be increased. We shall refer to Conjecture 5 as the Biconformability Conjecture.

CONJECTURE 5 (Biconformability Conjecture[1]). Let G be a bipartite graph having $\Delta(G) \geq 3 \big(|V(G)| + 1 \big) / 14$. Then G is Type 2 if and only if G contains an equibipartite subgraph H, with $\Delta(H) = \Delta(G)$, which is not biconformable.

---

[1] We must point out that we have recently been informed by Hung Lin Fu that he, B. L. Chen, J. G. Cheng, and K. C. Huang have found a family of counterexamples to the version of the Biconformability Conjecture presented here in the case where $\Delta(G) = n - 1$. It is too early to say whether or not these represent an isolated aberration, in the way that the Chen and Fu graphs seem to be an isolated aberration with respect to the original version of the Conformability Conjecture.

The Möbius ladder M_{14} shows that the number $3/14$ cannot be lowered.

In [**25**] Hilton showed that if G is equibipartite with $|V(G)| = 2n$, and if $\Delta(G) = n$, then G is Type 2 if and only if $e(\tilde{\tilde{G}}) + j(\tilde{\tilde{G}}) \leq n - 1$. From this it is easy to deduce the following proposition.

PROPOSITION 5.2. *Let G be a bipartite graph having bipartition (A, B), with $n = |A| \geq |B|$ and $n \geq 3$, and $\Delta(G) = n$. Then G is critical if and only if G is equibipartite and $e(\tilde{\tilde{G}}) + j(\tilde{\tilde{G}}) = n - 1$. Also $K_{2,2}$ is critical.*

PROOF. If G is equibipartite and $e(\tilde{\tilde{G}}) + j(\tilde{\tilde{G}}) = n - 1$, then $|V(G)| = 2n$, $\Delta(G) = n$, so G is Type 2. If e is any edge of G, then $G' = G\backslash e$ is also equibipartite and satisfies $|V(G')| = 2n$, $\Delta(G') = n$ and $e(\tilde{\tilde{G}}) + j(\tilde{\tilde{G}}) = n$. Therefore G is Type 1. Thus G is critical.

If G is critical then G is Type 2. First let us show that G must be equibipartite. Suppose $|A| > |B|$, then colour all vertices in B with colour c_{n+1}, and colour all the edges of G with the $\Delta(G) = n$ colours $c_1, c_2, \ldots, c_n$. Since $d_G(v) < n$ for every vertex $v \in A$, it follows that for each $v \in A$, at least one of the colours in $\{c_1, \ldots, c_n\}$ is not assigned to any edge incident with v. Colour each $v \in A$ with one of the colours missing at it. This gives a $\Delta + 1$ total colouring of G, which is a contradiction. Thus G is equibipartite.

If $e(\tilde{\tilde{G}}) + j(\tilde{\tilde{G}}) \geq n$, then G is Type 1, which is a contradiction. If $e(\tilde{\tilde{G}}) + j(\tilde{\tilde{G}}) \leq n - 2$, we show the existence of an edge in $E(G)$ whose removal does not increase $j(\tilde{\tilde{G}})$. In particular the deficit version of Hall's Theorem (see, for example [**7**, Corollary 4.3]) guarantees the existence of a subset $A' \subseteq A$ for which

$$n - j(\tilde{\tilde{G}}) = |A'| - \left|N_{\tilde{G}}(A')\right| = \max_{A'' \subseteq A}\left\{|A''| - \left|N_{\tilde{G}}(A'')\right|\right\}.$$

Choose A^* to have maximum size amongst all such sets A'. Clearly A^* includes all vertices of A which have degree zero in $\tilde{G}$ (otherwise we could enlarge A^* to A^{**} with $N_{\tilde{G}}(A^*) = N_{\tilde{G}}(A^{**})$, so that $|A^{**}| - |N_{\tilde{G}}(A^{**})| > |A^*| - |N_{\tilde{G}}(A^*)|$). Since $e(\tilde{\tilde{G}}) + j(\tilde{\tilde{G}}) \leq n - 2$, such vertices exist. If $N_{\tilde{G}}(A^*)$ is not empty, then since $\left|N_{\tilde{G}}(A'')\right| \geq |A''| - |A^*| + \left|N_{\tilde{G}}(A^*)\right|$ (for all $A'' \subseteq A^*$) it follows from the deficit version of Hall's Theorem that there is a partial matching of size $\left|N_{\tilde{G}}(A^*)\right|$ incident with the vertices of $N_{\tilde{G}}(A^*)$ in B. Choose e to be an edge of G incident with a vertex $x \in A$ of degree zero in $\tilde{G}$ and with some vertex of $N_{\tilde{G}}(A^*)$. The removal of this edge e from G does not increase $j(\tilde{\tilde{G}})$. If $N_{\tilde{G}}(A^*)$ is empty, consider the graph $\tilde{\tilde{G}}\backslash A^*$. Since $|A^*|$ has maximal size, it follows that, for all $A'' \subseteq A\backslash A^*$ with $A'' \neq \phi$, $\left|N_{\tilde{G}\backslash A^*}(A'')\right| > |A''|$. Therefore, by Hall's Theorem, there exists a matching of $A\backslash A^*$ into B. Since $|B| = n$ and $\tilde{\tilde{G}}$ has at most $n - 2$ edges, there is an edge $e \in E(G)$ joining a vertex in $A\backslash A^*$ to a vertex in B. Clearly, the removal of such an edge e from G does not increase $j(\tilde{\tilde{G}})$. Thus the graph $G' = G\backslash e$ satisfies $e(\tilde{\tilde{G}'}) + j(\tilde{\tilde{G}'}) \leq n - 1$. Since G is equibipartite, G' must be Type 2. Therefore G is not critical, a contradiction.

Thus if G is critical, then $e(\tilde{\tilde{G}}) + j(\tilde{\tilde{G}}) = n - 1$. □

The following proposition also supports the Biconformability Conjecture:

PROPOSITION 5.3. *Let G be a bipartite graph having bipartition (A, B), with $n = |A| \geq |B|$ and $n \geq 3$, and $\Delta(G) = n$. Then G is Type 2 if and only if G is equibipartite and not biconformable.*

PROOF. Lemma 5.1 proves that if G is equibipartite and not biconformable, then G must be Type 2.

Suppose that G is Type 2. Then the argument in the second paragraph of the proof of Proposition 5.2 shows that G is equibipartite, and then Hilton's result shows that $e(\tilde{\tilde{G}}) + j(\tilde{\tilde{G}}) \leq n - 1$. But if G is biconformable, then $\operatorname{def}(G) \geq \sum_{i=1}^{n+1} |a_i - b_i|$, where a_i and b_i are as in the definition of biconformability. Clearly

$$j(\tilde{\tilde{G}}) \geq n - \sum_{i=1}^{n+1} \max\{(a_i - b_i), 0\} \quad \text{and} \quad j(\tilde{\tilde{G}}) \geq n - \sum_{i=1}^{n+1} \max\{(b_i - a_i), 0\},$$

so

$$2j(\tilde{\tilde{G}}) \geq 2n - \sum_{i=1}^{n+1} |a_i - b_i| \geq 2n - \operatorname{def}(G).$$

Since $\Delta(G) = n$, it follows that $e(\tilde{\tilde{G}}) + j(\tilde{\tilde{G}}) \geq n$, a contradiction. Therefore G is not biconformable. ∎

The Biconformability Conjecture is also supported by Lemma 8.2. This lemma is interesting because the conditions $|V_{<\Delta}(A \backslash A_i)| \geq b_i - a_i$ and $|V_{<\Delta}(B \backslash B_i)| \geq a_i - b_i$ play a significant role in it; by contrast, they play no role in Proposition 5.2.

To end this section we prove a simple result that is similar in form to Theorem 3.4. There may be additional conditions which would allow a proof of the reverse implication.

THEOREM 5.4. *Let G be an equibipartite graph of even order $2n$ and maximum degree Δ. If $e(\tilde{\tilde{G}}) + j(\tilde{\tilde{G}}) \leq n(n - \Delta + 1) - 1$ then G is not biconformable.*

PROOF. We first note that the following inequalities are equivalent:

$$e(\tilde{\tilde{G}}) + j(\tilde{\tilde{G}}) \geq n(n - \Delta + 1), \quad 2j(\tilde{\tilde{G}}) \geq 2n + 2n(n - \Delta) - 2e(\tilde{\tilde{G}}),$$

$$2j(\tilde{\tilde{G}}) \geq 2n - \operatorname{def}(G).$$

If G is biconformable then

$$\operatorname{def}(G) \geq \sum_{i=1}^{\Delta(G)+1} |a_i - b_i|,$$

where a_i and b_i are as in the definition of biconformability. As shown in the proof of Proposition 5.3, this inequality implies that $2j(\tilde{\tilde{G}}) \geq 2n - \operatorname{def}(G)$, consequently $e(\tilde{\tilde{G}}) + j(\tilde{\tilde{G}}) \geq n(n - \Delta + 1)$. Thus if $e(\tilde{\tilde{G}}) + j(\tilde{\tilde{G}}) \leq n(n - \Delta + 1) - 1$ then G is not biconformable. ∎

6. Subgraphs of Critical Graphs

In this section we discuss a somewhat surprising divergence in the nature of critical edge-colourings and critical total colourings. Because there are two contrasting ideas of criticality, we restate the definitions: A graph G will be called *edge-critical* if G is connected and $\chi'(G \backslash e) < \chi'(G)$ for each edge $e \in E(G)$. Vizing's Adjacency Lemma (see [**21**]) implies that if G is edge-critical, then $\chi'(G) = \Delta(G) + 1$. In this section, a graph G will be called *totally critical* (earlier in this paper this was sometimes simply called critical) if G is connected and $\chi''(G\backslash e) < \chi''(G)$ for each edge $e \in E(G)$.

If G is an edge-critical graph, then G contains an edge-critical subgraph having maximum degree one less than $\Delta(G)$ (see [**21**]):

THEOREM 6.1. *Let G be an edge-critical graph having a maximum degree $\Delta \geq 3$. Then G contains an edge-critical subgraph H having maximum degree $\Delta - 1$.*

While this result is not particularly surprising, what is strange is that there is no analogous result for totally critical graphs. In particular, consider the graph 45 in our catalogue of critical graphs (see Section 8), which is simply the graph obtained by removing a C_7 and a C_3 that are vertex disjoint from the graph K_{10}. This graph has order 10, maximum degree 7 and is totally critical, but contains no totally critical subgraph of maximum degree 6 (it is easy to check that graph 45 does not contain any of graphs 30, 31 or 15 as a subgraph).

We will call graphs which are totally critical of maximum degree Δ, but which contain no totally critical subgraph of maximum degree $\Delta - 1$, *supercritical*. Thus graph number 45 in our catalogue is supercritical. However, most of the low order critical graphs G presented in the catalogue are as unlike supercritical graphs as it is possible to be, because they have a critical subgraph H of maximum degree $\Delta - 1$ which is spanning (in most cases $G\backslash E(H)$ consists of one or two P_2's (paths with two edges) and some independent edges). Notwithstanding this, if the Conformability Conjecture is true, then supercritical graphs are abundant, as the following result shows.

THEOREM 6.2. *If the Conformability Conjecture 2 is true, then any critical graph G with $|V(G)|$ even and $5 \leq |V(G)|/2 + 2 \leq \Delta(G) \leq 2|V(G)|/3 - 1$ is supercritical.*

PROOF. Let G be a critical graph of even order $2n$ satisfying $5 \leq |V(G)|/2 + 2 \leq \Delta(G) \leq 2|V(G)|/3 - 1$. Since $|V(G)| \geq 6$ it follows that $|V(G)| - 3 \geq 2|V(G)|/3 - 1 \geq \Delta$. Then it follows from Conjecture 2 and Theorem 3.6 that G is regular and G contains a K_{r_1,r_2} where r_1 and r_2 are odd and $r_1 + r_2 = |V(G)|$. Let this K_{r_1,r_2} have independent vertex sets X and Y.

Let H be a critical subgraph of G with $\Delta(H) = \Delta(G) - 1$. If $|V(H)| = |V(G)|$ then in view of the fact that, as above, H is also regular and contains a K_{s_1,s_2} where s_1 and s_2 are odd and $s_1 + s_2 = |V(H)| = |V(G)|$, it follows that $\{r_1, r_2\} = \{s_1, s_2\}$ and that the K_{s_1,s_2} is the original K_{r_1,r_2}. Thus H is formed from G by removing a 1-factor consisting of edges with both ends in X and edges with both ends in Y. But this is impossible as $|X|$ and $|Y|$ are both odd. Therefore $|V(H)| \neq |V(G)|$.

Consider the possibility that $|V(H)|$ is even. Then H is a subgraph of some graph J obtained from G by removing two vertices. Since $\Delta(G) \geq |V(G)|/2 + 2$, $\delta(J) \leq \Delta - 2$. Consider a graph J^*, containing H as a subgraph, obtained from J by removing a minimal set of edges of $E(J)\backslash E(H)$ so that $\Delta(J^*) = \Delta(H)$. Then $\Delta(J^*) = \Delta(G) - 1$ and $\delta(J^*) \leq \Delta(G) - 2$, so J^* is not regular. Since $\Delta(J^*) = \Delta(G) - 1 \geq |V(G)|/2 + 1 = |V(J^*)|/2 + 2$, it follows from Theorem 3.6 that either J^* is conformable or that $\Delta(J^*) > 3|V(J^*)|/4 - 1$. In the latter case we have

$$\tfrac{2}{3}|V(G)| - 1 \geq \Delta(G) = \Delta(J^*) + 1 > \tfrac{3}{4}|V(J^*)|$$

so that $2(2n)/3 - 1 > 3(2n-2)/4$, i.e. $6 > 2n$, which is not the case, as we assume that $5 \leq n + 2$. It follows that J^* is conformable. Therefore $V(H) \neq V(J^*)$. Thus H is a spanning subgraph of a graph J^{**} obtained from J^* by removing an even number of vertices. Therefore $|V(J^{**})| \leq |V(G)| - 4$. Recall that G contains a K_{r_1,r_2} with independent vertex sets X and Y, and that, by Theorem 3.6, $\min(r_1,r_2) \geq 2n - \Delta(G)$. Since $\Delta(J^{**}) = \Delta(G) - 1$ and $|V(J^{**})| \leq |V(G)| - 4$, J^{**} is a subgraph of a graph obtained from G by removing at least three vertices of X or three vertices of Y. Therefore

$$\text{def}(H) \geq \text{def}(J^{**}) \geq 2\min(r_1,r_2) \geq 2(2n - \Delta(G))$$
$$\geq 2(2n - (\tfrac{2}{3}(2n) - 1)) = \tfrac{2}{3}(2n) + 2 > \Delta(H) + 1,$$

so H is conformable, a contradiction.

Thus $|V(H)|$ is odd. Then one of $|X \cap V(H)|$ and $|Y \cap V(H)|$ is even, the other odd; we may suppose $|X \cap V(H)|$ is even. Now let us consider possible conformable vertex colourings of H.

We initially colour the vertices of $Y \cap V(H)$ with one colour and colour the vertices of $X \cap V(H)$ with a further $|X \cap V(H)|$ colours. Then $|X \cap V(H)| + 1$ colours are each used on an odd number of vertices. Since

$$1 + |X \cap V(H)| \leq 1 + \max\{r_1 - 1, r_2 - 1\} \leq \Delta(G) = \Delta(H) + 1,$$

at most $1 + \Delta(H)$ colours are actually used initially. If $1 + |X \cap V(H)| < \Delta(H)$ we increase the number of colours occurring on an odd number of vertices by recolouring $\Delta(H) - |X \cap V(H)|$ vertices of $Y \cap V(H)$ with a further $\Delta(H) - |X \cap V(H)|$ colours. Then $\Delta(H)$ vertices form $\Delta(H)$ singleton colour classes, so $V(H)$ has either $\Delta(H) + 1$ odd colour classes and so no even colour classes, and thus has a conformable vertex-colouring, or has $\Delta(H)$ singleton colour classes and one even colour class. If H is not regular, this would also be a conformable vertex-colouring. If H is regular, then, since $|V(H)|$ is odd, $\Delta(H)$ is even. But then $\Delta(H)$ odd numbers and one even number sum to an even number, a contradiction. Thus H is not regular.

This proves Theorem 6.2. □

Now consider the following family of graphs: Let p be any odd integer such that $p \geq 7$. Define S_p to be the graph obtained by adding edges joining vertices of the independent set of size p in a $K_{p-4,p}$, so that the new edges form the square of a p-cycle on these vertices. The graph S_p is regular of degree p. [If p is even, the corresponding graph is conformable, by Theorem 3.6.]

LEMMA 6.3. *If $p \geq 7$ and is odd, then the graph S_p is totally critical.*

PROOF. For convenience, let $V(S_p) = X \cup Y$ where $X = \{x_1, x_2, \ldots, x_{p-4}\}$ and $Y = \{y_1, y_2, \ldots, y_p\}$ and let $E(S_p)$ comprise all edges of the form $x_i y_j$, for $1 \leq i \leq p - 4$ and $1 \leq j \leq p$, and all edges of the form $y_i y_{i+1}$ and $y_i y_{i+2}$, where $1 \leq i \leq p$ and addition is modulo p.

Note that there are essentially two types of edges in $E(S_p)$; namely, those joining vertices in X to vertices in Y and those joining two vertices in Y. Since S_p is nonconformable and thus Type 2, to show that S_p is totally critical we need only present $(p+1)$-total colourings of the graph $S_p - e$ for each type of edge e.

To begin the proof we present an assignment of colours to the vertices and edges of S_p. This assignment of colours will then be modified to provide a $(p+1)$-total colouring of $S_p - e$ for each type of edge e. In our original assignment of colours, some vertices will be assigned more than one colour while some edges may not be assigned any colour. Furthermore adjacent elements may be assigned the same colour.

For $i = 1, 2, \ldots, \lfloor p/2 \rfloor$, assign colour c_i to the edges

$$x_1 y_{2i-1}, x_2 y_{2i}, x_3 y_{2i+1}, \ldots, x_{p-4} y_{2i+p-6} \quad \text{and} \quad y_{2i+p-5} y_{2i+p-3}, y_{2i+p-4} y_{2i+p-2}.$$

For $j = 1, 2, \ldots, \lfloor p/2 \rfloor$, assign colour $c_{\lfloor p/2 \rfloor + j}$ to the edges

$$x_1 y_{2j}, x_2 y_{2j+1}, x_3 y_{2j+2}, \ldots, x_{p-4} y_{2j+p-5} \quad \text{and} \quad y_{2j+p-3} y_{2j+p-2},$$

and to the vertices

$$y_{2j+p-4} \quad \text{and} \quad y_{2j+p-1}.$$

Assign colour c_p to the edges

$$x_1 y_p, x_2 y_1, x_3 y_2, \ldots, x_{p-4} y_{p-5} \quad \text{and} \quad y_{p-4} y_{p-2}$$

and to the vertices

$$y_{p-3} \quad \text{and} \quad y_{p-1}.$$

Finally assign the colour c_{p+1} to the edges

$$y_{p-2} y_{p-1}, y_p y_1, y_2 y_3, \ldots, y_{p-5} y_{p-4}$$

and to the vertices in X.

It is worth noting that the set of edges assigned the colour c_i is a perfect matching in S_p for $i = 1, 2, \ldots, \lfloor p/2 \rfloor$. For $j = \lfloor p/2 \rfloor + 1, \lfloor p/2 \rfloor + 2, \ldots, 2\lfloor p/2 \rfloor$, the set of vertices and edges assigned the colour c_j comprises two vertices and a matching containing edges incident with every vertex in $V(S_p)$ except for the two vertices coloured c_j. It follows that if $1 \le i \le \lfloor p/2 \rfloor$ and $\lfloor p/2 \rfloor + 1 \le j \le 2\lfloor p/2 \rfloor$, then the graph $G_{i,j}$ induced by the edges coloured c_i or c_j consists of a path joining the two vertices coloured c_j and, possibly, a number of even cycles.

The $p+1$ colours assigned to the vertices and edges of G would provide a proper total colouring of G except for the following problems:

(i) The edge $y_{p-3} y_{p-2}$ is not assigned a colour.
(ii) The vertex y_{p-2} is assigned two colours, $c_{\lfloor p/2 \rfloor + 1}$ and $c_{2\lfloor p/2 \rfloor}$.
(iii) The vertices y_{p-3} and y_{p-1} are given the same colour c_p.

We now show how to adapt the given assignment of colours to provide a $(p+1)$-total colouring of $S_p - e$ for the two possible choices of e.

CASE 1. The edge e joins two vertices in Y. Without loss of generality $e = y_{p-3} y_{p-1}$.

In this case (iii) is no longer a problem. Furthermore, we can resolve (ii) by choosing to colour the vertex y_{p-2} with $c_{\lfloor p/2 \rfloor + 1}$.

To resolve problem (i), we consider the component of $G_{1, 2\lfloor p/2 \rfloor} \backslash \{e\}$ containing the vertex y_{p-3}. Since $y_{p-3} y_{p-1}$ was originally assigned the colour c_1, this component must be a path which begins at y_{p-3}. It is easy to see that the sequence of edges in the path is

$$y_{p-3} y_{p-4}, y_{p-4} x_{p-4}, x_{p-4} y_{p-6}, y_{p-6} x_{p-6}, x_{p-6} y_{p-8}, \ldots, x_3 y_1, y_1 x_1, x_1 y_{p-1}.$$

Thus the path ends at y_{p-1}.

We swop the colours, c_1 and $c_{2\lfloor p/2 \rfloor}$, assigned to the edges in this path. The colour $c_{2\lfloor p/2 \rfloor}$ is, therefore, missing at y_{p-3}. Furthermore, since we chose to colour y_{p-2} with $c_{\lfloor p/2 \rfloor + 1}$, the colour $c_{2\lfloor p/2 \rfloor}$ is also missing at y_{p-2}. Thus we can assign $c_{2\lfloor p/2 \rfloor}$ to the uncoloured edge $y_{p-3}y_{p-2}$ and the resulting assignment of colours is a proper $(p+1)$-total colouring of $S_p - e$.

CASE 2. The edge e joins a vertex in X to a vertex in Y. Without loss of generality $e = x_{p-5}y_{p-3}$.

In this case the edge e was originally coloured c_2. To resolve (ii), we choose to colour the vertex y_{p-2} with $c_{2\lfloor p/2 \rfloor}$.

To resolve problems (i) and (iii), we consider the component of $G_{2,2\lfloor p/2 \rfloor - 1} \setminus \{e\}$ containing the vertex y_{p-3}. Since $y_{p-3}x_{p-5}$ was originally assigned the colour c_2, this component must be a path, P say, which begins at y_{p-3}. We shall show that the path P ends at y_{p-4} if for some positive integer k, $p = 1 + 6k$ or $p = 3 + 6k$, while P ends at y_{p-7} if $p = 5 + 6k$. We will then show how to reassign the colours to obtain a $(p+1)$-total colouring of $S_p - e$ in each case.

Initially note that the edge-coloured c_2 incident with a vertex x_ℓ (for $1 \leq \ell \leq p - 4$) is

$$x_\ell y_{\ell+2}$$

while the edge-coloured $c_{2\lfloor p/2 \rfloor - 1}$ incident with a vertex y_ℓ (for $1 \leq \ell \leq p - 8$) is

$$y_\ell x_{\ell+4}.$$

CASE 2.1 $(p = 1 + 6k)$. The sequence of edges in P is

$$y_{p-3}x_1, x_1y_3, y_3x_7, x_7y_9, \ldots, x_{1+6(k-1)}y_{3+6(k-1)} = x_{p-6}y_{p-4}.$$

So in this case the path P ends at y_{p-4}.

We will defer a description of the reassignment of colours until the end of Case 2.2 as the reassignment is the same in each of these cases.

CASE 2.2 $(p = 3 + 6k)$. The description of the path P is more complex in this case. It is:

$$y_{p-3}x_1, x_1y_3, y_3x_7, \ldots, y_{p-12}x_{p-8}, x_{p-8}y_{p-6}, y_{p-6}y_{p-5}, y_{p-5}x_{p-7},$$

$$x_{p-7}y_{p-11}, y_{p-11}x_{p-13}, \ldots, x_8y_4, y_4x_2, x_2y_{p-2}, y_{p-2}x_{p-4}, \ldots, x_{11}y_7,$$

$$y_7x_5, x_5y_1, y_1y_{p-1}, y_{p-1}x_3, x_3y_5, \ldots, x_{p-12}y_{p-10}, y_{p-10}x_{p-6}, x_{p-6}y_{p-4}.$$

Thus the path P ends at y_{p-4} in this case.

To reassign the colours, we use the following procedure:

(a) Recolour the vertex y_{p-3} (originally coloured c_p) with $c_{2\lfloor p/2 \rfloor - 1}$.
(b) Swop the colours c_2 and $c_{2\lfloor p/2 \rfloor - 1}$ assigned to the edges of the path P.
(c) Recolour the vertex y_{p-4} (originally coloured $c_{2\lfloor p/2 \rfloor - 1}$) with colour c_p.
(d) Recolour the edge $y_{p-4}y_{p-2}$ (originally coloured c_p) with colour $c_{\lfloor p/2 \rfloor + 1}$.
(e) Recolour the edge $x_{p-5}y_{p-4}$ (originally coloured $c_{\lfloor p/2 \rfloor + 1}$) with colour c_2.
(f) Colour the uncoloured edge $y_{p-3}y_{p-2}$ with c_p.

CASE 2.3 $(p = 5 + 6k)$. In this case the path P is fairly easy to describe but ends at y_{p-7} rather than y_{p-4} so the previous recolouring procedure does not work.

To help the description of the recolouring, consider the graph $G_{2,p} - \{y_{p-4}y_{p-2}\}$. Let Q be the path in this graph starting at y_{p-1}. We shall show that Q ends at y_{p-4}.

Initially, note that the edge-coloured c_p incident with a vertex y_ℓ (for $1 \le \ell \le p-5$) is

$$y_\ell x_{\ell+1}.$$

Thus the sequence of edges in the path Q is

$$y_{p-1}y_1, y_1x_2, x_2y_4, y_4x_5, \ldots, x_{5+6(k-1)}y_{1+6k} = x_{p-6}y_{p-4}.$$

To reassign the colours, we use the following procedure:

(a) Recolour the vertex y_{p-3} (originally coloured c_p) with $c_{\lfloor p/2 \rfloor + 1}$ (this resolves problem (iii)).
(b) Recolour the edge $y_{p-3}x_{p-4}$ (originally coloured $c_{\lfloor p/2 \rfloor + 1}$) with colour c_2.
(c) Recolour the edge $x_{p-4}y_{p-2}$ (originally coloured c_2) with colour $c_{\lfloor p/2 \rfloor + 1}$.
(d) Recolour the edge $y_{p-3}y_{p-2}$ (originally uncoloured) with colour c_p (this resolves problem (i)).
(e) Recolour the edge $y_{p-2}y_{p-4}$ (originally coloured c_p) with colour c_2.
(f) Swop the colours c_2 and c_p assigned to the edges of the path Q.
(g) Recolour the vertex y_{p-1} (originally coloured c_p) with colour c_2.

Since we obtain a $(p+1)$-total colouring of $S_p - e$ in each case, it follows that S_p is totally critical. $\qquad\square$

In view of Theorem 6.2 and Lemma 6.3 we can sum up the position as regards the graphs S_p as follows:

THEOREM 6.4. *For $p \ge 7$, p odd, the family of graphs S_p is an infinite family of totally critical graphs. If the Conformability Conjecture is correct, it is an infinite family of supercritical graphs.*

It would be interesting to know if there are any supercritical graphs G with $|V(G)|$ even and $\Delta(G) < |V(G)|/2$, or with $|V(G)|$ odd.

7. Open Problems

In this section we mention a few interesting questions which arose from the research done for this paper.

The following question is intriguing. The analogous claim for edge colourings is known to be false.

QUESTION 1. Is the following claim true?
Let r be any fixed positive integer. The number of critical graphs having exactly r vertices of maximum degree is finite.

A second, even stronger, claim is the following. It is easy to see that the truth of the claim in Question 2 would imply the truth of the claim in Question 1.

QUESTION 2. Is the following claim true?
Let G be a critical graph. Then the number of vertices of maximum degree is at least $|V(G)|/2$.

Let $C_n(r)$ be the number of critical graphs of order n with r vertices of maximum degree. Values for $C_n(r)$, obtained from our catalogue, are given in the table below:

n \ r	1	2	3	4	5	6	7	8	9	10	Total
2	0	1									1
3	0	0	0								0
4	0	0	0	2							2
5	0	0	0	0	1						1
6	0	0	0	2	1	0					3
7	0	0	0	0	0	1	2				
8	0	0	0	0	2	0	1	3			6
9	0	0	0	0	0	1	4	6	4		15
10	0	0	0	0	0	4	1	1	1	12	19
Total	0	1	0	4	4	6	8	10	5	12	50

8. A Table of Critical Graphs of Order ≤ 10

In this section we present a catalogue of the critical graphs having order at most ten. Associated with each graph is a triple (n, Δ, m), where n is the order, Δ is the maximum degree and m is the size of the graph. The graphs are listed in increasing order of n, then Δ, then m. The catalogue is complete up to and including the graphs of order eight, and is believed to be complete up to and including the graphs of order ten.

To conclude, we try to give short reasons why the critical graphs listed in the table are Type 2. Certain graphs will fall into more than one of the categories (for example graph 2 is both a cycle and a Chen and Fu graph). We have, however, simply listed such graphs in the first categories which apply to them.

Nonconformability. We have shown in Proposition 2.1, that if a graph is nonconformable, then it is Type 2. The following graphs in the catalogue are nonconformable 1, 3, 7, 10, 14, 16, 27, 28, 29, 30, 31, 44, 45, 46, 48, 49 and 50. Moreover some are critical because they satisfy Theorem 3.2 or Propositions 4.2 or Theorem th4.6, and that the others are critical is predicted by Conjecture 2.

Cycles. It is well known (see for example [4]) that a cycle which has an order which is not a multiple of 3 is critical. The following graphs fall in this class: 2, 4, 8, 11 and 32.

Graphs having maximum degree three. These graphs are presented in an earlier paper by Hamilton and Hilton [23]. They will not be discussed in detail in this section. The following graphs are in this class: 9,12,17,18, 19, 20 and 33 to 40.

Chen and Fu graphs. Chen and Fu [9] showed that any graph obtained from an odd order complete graph by subdividing one edge is critical even though it is conformable. This class contains graphs numbers 6, 15 and 47.

Nonbiconformability. In [25] Hilton showed that if J is a subgraph of $K_{n,n}$ with $e = |E(J)|$ and j independent edges in J, then $G = K_{n,n} \backslash E(J)$ has total chromatic number $n + 2$ if and only if $e + j \leq n - 1$. This explains graphs numbers 5, 13, 42 and 43. In Lemma 8.2 below it is explained why all members of a general class of graphs, which includes graph number 41, are not biconformable, and why they are critical.

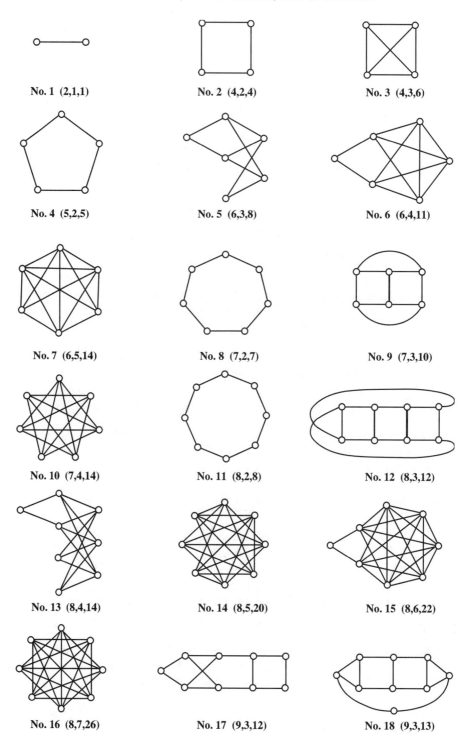

No. 1 (2,1,1)

No. 2 (4,2,4)

No. 3 (4,3,6)

No. 4 (5,2,5)

No. 5 (6,3,8)

No. 6 (6,4,11)

No. 7 (6,5,14)

No. 8 (7,2,7)

No. 9 (7,3,10)

No. 10 (7,4,14)

No. 11 (8,2,8)

No. 12 (8,3,12)

No. 13 (8,4,14)

No. 14 (8,5,20)

No. 15 (8,6,22)

No. 16 (8,7,26)

No. 17 (9,3,12)

No. 18 (9,3,13)

FIGURE 1. Catalogue of critical graphs.

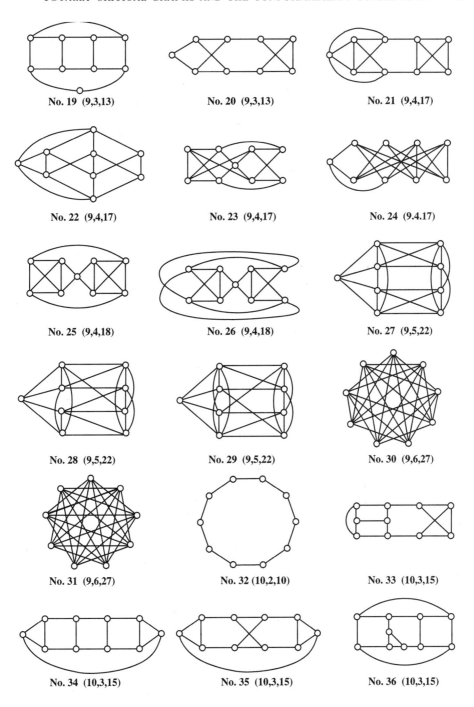

No. 19 (9,3,13)   No. 20 (9,3,13)   No. 21 (9,4,17)

No. 22 (9,4,17)   No. 23 (9,4,17)   No. 24 (9.4.17)

No. 25 (9,4,18)   No. 26 (9,4,18)   No. 27 (9,5,22)

No. 28 (9,5,22)   No. 29 (9,5,22)   No. 30 (9,6,27)

No. 31 (9,6,27)   No. 32 (10,2,10)   No. 33 (10,3,15)

No. 34 (10,3,15)   No. 35 (10,3,15)   No. 36 (10,3,15)

FIGURE 1. (continued)

No. 37 (10,3,15)   No. 38 (10,3,15)   No. 39 (10,3,15)

No. 40 (10,3,15)   No. 41 (10,4,18)   No. 42 (10,5,22)

No. 43 (10,5,23)   No. 44 (10,7,35)   No. 45 (10,7,35)

No. 46 (10,7,35)   No. 47 (10,8,37)   No. 48 (10,9,42)

No. 49 (10,9,42)   No. 50 (10,9,43)

FIGURE 1. (continued)

"Same and different" graphs. A number of graphs are Type 2 because they have the following structure: The graph G possesses a two edge-cutset $\{e_1, e_2\}$ and two (connected) subgraphs, G_1 and G_2 say, such that $E(G) = E(G_1) \cup E(G_2)$, $E(G_1) \cap E(G_2) = \{e_1, e_2\}$ and $V(G_1) \cap V(G_2)$ comprises the endvertices of e_1 and e_2. Furthermore, any $\Delta + 1$-total colouring of G_1 assigns the same colour to both e_1 and e_2, while any $\Delta + 1$-total colouring of G_2 assigns different colours to e_1 and e_2. Since a total colouring of G induces a total colouring of G_1 and a total colouring of G_2, it follows that G must be Type 2. Graph number 21 has this structure. A number of graphs having maximum degree three also fall into this class.

Further critical graphs. The graphs numbered 22, 23, 24, 25 and 26 are not included in any of the above categories. All these graphs seem to fit two general patterns: Graphs 23, 25 and 26 are based on the idea of taking the union of two complete even order graphs and an isolated vertex and matching some corresponding vertices in the complete graphs. Graphs 22 and 24 involve complete equibipartite graphs from which a matching has been removed and to which one further vertex has been added; these seem to be similar to graph number 41.

The following lemma explains why graph number 25 is Type 2 (graphs numbers 23 and 26 appear to be close variants of graph number 25, so we can expect to find explanations why these graphs are Type 2 based on similar ideas). We also show that such graphs are critical.

LEMMA 8.1. *Let G be a graph having order $2n + 1$, where $n \geq 4$ and n is even, formed from two vertex disjoint K_n's by adding a matching of size $n - 2$ joining $n - 2$ vertices from each K_n, and by joining the remaining two vertices in each K_n to a new vertex. Then G is critical.*

PROOF. Suppose that G is Type 1 and consider a $\Delta(G) + 1$ total colouring of G. Since $\Delta(G) = n$, at most $n + 1$ colours can be used in the colouring. Each of the $n + 1$ colours must be present at each of the vertices of degree n. Suppose, without loss of generality, that the vertices of one of the subgraphs isomorphic to K_n, subgraph H_1 say, are assigned colours $c_1, c_2, \ldots, c_n$. By counting, it follows that in H_1 there are $n/2$ edges coloured c_{n+1}, and $n/2 - 1$ edges coloured with each of $c_1, \ldots, c_n$. Consequently each of the n edges of G, which are incident with exactly one of the vertices in H_1, must be assigned a distinct colour from $c_1, c_2, \ldots, c_n$. Repeating this argument with the second complete subgraph, H_2 say, we deduce that each of the n edges of G, which are incident with exactly one of the vertices in H_2, must be assigned a distinct colour from $c_1, c_2, \ldots, c_n, c_{n+1}$. Since $n - 2$ of the colours from $c_1, c_2, \ldots, c_n, c_{n+1}$ are assigned to edges of the matching joining H_1 and H_2, it follows that the remaining four edges (each of which is incident with the vertex not in $V(H_1) \cup V(H_2)$) must be assigned one of three remaining colours. Thus G cannot be Type 1.

To show that G is critical, we must show that the removal of any edge gives a Type 1 graph. In this section we may assume that $n \geq 6$ as the case $n = 4$ is already known to be critical as it is graph number 25 in our catalogue. Notice that there are five types of nonisomorphic edges in $E(G)$.

For convenience let H_1 be one of the subgraphs of G isomorphic to K_n and let H_1 have vertices $x_1, x_2, \ldots, x_n$. Let H_2 be the other subgraph of G isomorphic to K_n and let H_2 have vertices $y_1, y_2, \ldots, y_n$. Let z be the vertex of degree four. Let $x_1 y_1, x_2 y_2, \ldots, x_{n-2} y_{n-2}$ be the edges joining vertices in H_1 to vertices in H_2 and let $x_{n-1} z, x_n z, y_{n-1} z, y_n z$ be the edges joining vertices in H_1 or H_2 to z.

Consider $G_1 = G - y_{n-1} z$. Colour the vertices of H_1 with colours $c_1, c_2, \ldots, c_n$ and the vertices of H_2 with $c_1, c_2, \ldots, c_{n-1}$ and c_{n+1}. It is possible (provided the vertex-colourings are chosen appropriately) to colour the edges of H_1 and H_2 with $c_1, c_2, \ldots, c_{n+1}$ in such a way that colour c_i is missing at x_i for $i = 1, 2, \ldots, n$, colour c_j is missing at y_j for $j = 1, 2, \ldots, n - 1$, and colour c_{n+1} is missing at y_n.

Then colouring x_iy_i with c_i for $i = 1, 2, \ldots, n-2$, $x_{n-1}z$ with c_{n-1}, x_nz with c_n and y_nz with c_{n+1} gives a total colouring of G_1 using $\Delta + 1 = n + 1$ colours.

The other cases are all similar to this. If $G_2 = G - x_{n-2}y_{n-2}$, colour H_1 as above, and modify the colouring of H_2 so that all the missing colours are as before, except that colours c_{n-1}, c_{n-2} and c_{n+1} are missing from y_{n-2}, y_{n-1} and y_n, respectively. Then a total colouring of G_2 is obtained in an obvious way.

If $G_3 = G - y_{n-1}y_n$, colour H_1 as above, and colour all of H_2 with the same colours missing at the vertices as in G_1, but with $y_{n-1}y_n$ coloured c_n, and y_{n-1} coloured c_1. Then remove the edge $y_{n-1}y_n$, recolour y_{n-1} with c_n, and extend the colourings of H_1 and $H_2\backslash y_{n-1}y_n$ to all of G_3 in an obvious way, with zy_{n-1} and zy_n coloured c_1 and c_{n+1} respectively.

If $G_4 = G - y_{n-2}y_{n-1}$, colour H_1 as above, and colour all of H_2 with the same colours missing at the vertices as in G_1, but with $y_{n-2}y_{n-1}$ coloured c_n, and y_{n-1} coloured c_1. Then remove the edge $y_{n-2}y_{n-1}$, recolour y_{n-1} with c_n, and extend the colourings of H_1 and $H_2\backslash y_{n-2}y_{n-1}$ to all of G_4 in an obvious way, with zy_{n-1} and zy_n coloured c_1 and c_{n+1} respectively.

Finally, if $G_5 = G - y_{n-3}y_{n-2}$, colour H_1 as above, and colour all of H_2 with the same colours missing at the vertices as in G_1, but with $y_{n-3}y_{n-2}$, $y_{n-1}y_{n-2}$, y_1y_{n-1} and y_1y_{n-3} coloured c_n, c_1, c_n and c_{n-1}, respectively. This is possible since $n \geq 6$. (There is more than one way of seeing this; one way is to observe that a total colouring of K_{n+1}, with n even, with $n + 1$ colours is equivalent to an edge-colouring of K_{n+2} with the same set of colours, and then to use the fact, shown in [1, Corollary 4.3.9], that any edge colouring of $K_{n+2/2}$ with $n + 1$ colours can be extended to an edge-colouring of K_n with the same set of colours.) Then remove the edge $y_{n-3}y_{n-2}$, recolour y_1y_{n-1} with c_{n-1} and $y_{n-2}y_{n-1}$ with c_n. Then extend the colourings of H_1 and $H_2\backslash y_{n-3}y_{n-2}$ to all of G_5 in an obvious way, with zy_{n-1} and zy_n coloured c_1 and c_{n-1}, respectively.

Since the removal of any other edge leaves a graph isomorphic to one of the five graphs considered above, it follows that G is critical. □

The following lemma explains why graph number 41 is not biconformable, and also why it is critical. Lemma 8.2 is the best approach we have so far to an extension of Proposition 5.2 to bipartite graphs with bipartition (A, B), with $|A| \geq |B|$, such that $|A| > \Delta(G)$.

LEMMA 8.2. *Let G be a graph of order $2n+2$, where $n \geq 4$, obtained by removing $n - 2$ independent edges $x_1y_1, x_2y_2, \ldots, x_{n-2}y_{n-2}$ from a $K_{n,n}$, and joining a new vertex y^* to $x_1, x_2, \ldots, x_{n-2}$ and a second new vertex x^* to $y_1, y_2, \ldots, y_{n-2}$. Then G is not biconformable and is critical.*

PROOF. Let the independent vertex-sets of G be $A = \{x^*, x_1, x_2, \ldots, x_n\}$ and $B = \{y^*, y_1, y_2, \ldots, y_n\}$.

We show first that G is not biconformable. Suppose G has a vertex-colouring with colours $\{1, 2, \ldots\}$. Let A_i, B_i denote the set of vertices of A, B, respectively, coloured i. Suppose the vertex-colouring satisfies the two conditions

$$\left|V_{<\Delta}(A\backslash A_i)\right| \geq b_i - a_i \quad (\forall i)$$

and

$$\left|V_{<\Delta}(B\backslash B_i)\right| \geq a_i - b_i \quad (\forall i).$$

If x_{n-1} or x_n is coloured c, then the only vertex in B which can be coloured c is y^*. If y^* is not coloured c, then $b_c = 0$ so

$$\left|V_{<\Delta}(B\backslash B_c)\right| = 1 \geq a_c - b_c = a_c - 0,$$

so $a_c = 1$, and so no other vertex in A is coloured c. If y^* is coloured c, then $b_c = 1$ and so

$$\left|V_{<\Delta}(B\backslash B_c)\right| = 0 \geq a_c - b_c = a_c - 1;$$

therefore $a_c = 1$, so no other vertex in A is coloured c. Thus if x_{n-1} or x_n is coloured c then the only other vertex which can be coloured c is y^*. Similarly, if y_{n-1} or y_n is coloured c, the only other vertex which can be coloured c is x^*.

If x_j is coloured c for some $j \in \{1, 2, \ldots, n-2\}$, the only vertex of B which can be coloured c is y_j. If y_j is not coloured c, then $b_c = 0$, so

$$\left|V_{<\Delta}(B\backslash B_c)\right| = 1 \geq a_c - b_c = a_c - 0,$$

so $a_c = 1$, and so no other vertex in A is coloured c. If y_j is coloured c, then again no other vertex in A is coloured c. Thus if x_j is coloured c for some $j \in \{1, 2, \ldots, n-2\}$, then only one other vertex can be coloured c, namely y_j.

Thus $n - 2$ colours are needed for $x_1, x_2, \ldots, x_{n-2}$ and $y_1, y_2, \ldots, y_{n-2}$, and a further four colours are needed for x_{n-1}, x_n, y_{n-1} and y_n. Thus at least $n + 2$ colours are needed altogether, and so G is not biconformable.

By Lemma 5.1, it follows that G is Type 2.

Next we show that G is critical. We give a formal proof which is valid only for $n \geq 10$. The remaining cases were checked on a computer, but we do not give the details here. To show that G is critical we have to show that $G\backslash e$ can be totally coloured with $n + 1$ colours for each edge e of G. There are exactly four nonisomorphic graphs which can be obtained from G by removing a single edge; they can be obtained by removing the edges $x_{n-2}y^*$, x_1y_2, $x_{n-2}y_{n-1}$ and $x_{n-1}y_{n-1}$ respectively.

Let G^* be the $K_{n+1,n+1}$ obtained from G by adding the edges x^*y_{n-1}, x^*y_n, y^*x_{n-1}, y^*x_n, x^*y^* and x_iy_j for $1 \leq i \leq n - 2$. In each case we shall give G^* an appropriate edge-colouring with colours $c_1, c_2, \ldots, c_{n+1}$, and then derive from it a total colouring of $G\backslash e$.

First consider the graph $G_1 = G - x_{n-2}y^*$. In G^* colour the edges x^*y^*, x^*y_{n-1}, x^*y_n, y^*x_{n-2}, y^*x_{n-1}, y^*x_n, $x_{n-2}y_{n-2}$ with colours c_{n-2}, c_1, c_{n+1}, c_1, c_n, c_{n-1}, c_{n-2}, respectively. Now extend this partial edge-colouring of G^* to an edge-colouring of all of G^* with colours $c_1, \ldots, c_{n+1}$ in such a way that the edge x_iy_i, for $1 \leq i \leq n - 3$, receives colour c_i. Bearing in mind the correspondence between an edge-coloured $K_{n+1,n+1}$ and a latin square of side $n + 1$ [where the two maximal independent vertex-sets in the $K_{n+1,n+1}$ correspond to the row and column labels of the latin square, and the coloured edges correspond to the symbols in the latin square], it follows from [2] that such an extension of the partial edge-colouring exists for all $n \geq 8$. We can now obtain the required total colouring of G_1 as follows. We remove the edges of G^* which are not in G_1. All the edges of G_1 are coloured. We recolour the edge $x_{n-2}y_{n-1}$ with colour c_1, and we colour the vertices x^*, y^*, x_{n-2}, x_{n-1}, x_n, y_{n-2}, y_{n-1}, y_n with colours c_{n+1}, c_{n-2}, c_{n-2}, c_n, c_{n-1}, c_{n-2}, c_{n+1}, c_{n+1} respectively, and colour x_i and y_i with colour c_i for $1 \leq i \leq n - 3$. It is easy to see that this yields a total colouring of G_1.

Next consider the graph $G_2 = G - x_1y_2$. In G^* colour the edges x_1y_1, x_2y_2, x_1y_2, x_1y_{n-1}, x^*y_{n-1}, x^*y_n, y^*x_{n-1}, y^*x_n with colours c_1, c_2, c_3, c_{n+1}, c_3, c_{n+1}, c_{n-1}, c_n, respectively and, for $3 \leq i \leq n - 2$, colour the edge x_iy_i with colour c_i. Then extend this partial edge-colouring of G^* to an edge-colouring of all of G^* with colours $c_1, \ldots, c_{n+1}$. For $n \geq 10$ this can be done by [2]. Now we obtain the required total colouring of G_2 as follows. We remove the edges of G^* which are not in G_2. We recolour the edge x_1y_{n-1} with colour c_3. Then we colour x^*, y^*, x_1, x_2, x_{n-1}, x_n, y_1, y_2, y_{n-1}, y_n with colours c_{n+1}, c_n, c_1, c_2, c_{n-1}, c_n, c_1, c_2, c_{n+1}, c_{n+1} respectively, and colour x_i and y_i with c_i for $3 \leq i \leq n - 2$. It is easy to check that this yields a total colouring of G_2.

Next consider the graph $G_3 = G - x_{n-2}y_{n-1}$. In G^* colour the edges x^*y_n, y^*x_{n-1}, y^*x_n, $x_{n-2}y_{n-2}$, $x_{n-2}y_{n-1}$ with colours c_{n+1}, c_{n-1}, c_n, c_{n-2}, c_{n+1} respectively, and for $1 \leq i \leq n - 3$, colour the edge x_iy_i with c_i. Then extend this partial edge-colouring of G^* to an edge-colouring of all of G^*. This can be done for all $n \geq 6$ by [2]. Now obtain a total colouring of G_3 as follows. Remove the edges of G^* which are not in G_3. Then colour vertices x^*, y^*, x_{n-1}, x_n, y_{n-1}, y_n with colours c_{n+1}, c_n, c_{n-1}, c_n, c_{n+1}, c_{n+1} respectively, and colour x_i and y_i with c_i for $1 \leq i \leq n - 2$. This yields a total colouring of G_3.

Finally consider the graph $G_4 = G - x_{n-1}y_{n-1}$. In G^* colour the edges x^*y_n, y^*x_n, $x_{n-1}y_{n-1}$ with colours c_{n+1}, c_n, c_{n-1} respectively, and, for $1 \leq i \leq n - 2$, colour the edge x_iy_i with c_i. Then extend this partial edge-colouring to an edge-colouring of all of G. This can be done for $n \geq 6$ by [2]. Now obtain the required total colouring of G_4 as follows. Remove the edges of G^* which are not in G_4. Then colour the vertices x^*, y^*, x_{n-1}, x_n, y_{n-1}, y_n with c_{n+1}, c_n, c_{n-1}, c_n, c_{n-1}, c_{n+1} respectively, and, for $1 \leq i \leq n - 2$, colour x_i and y_j with colour c_i. This yields a total colouring of G_4.

Since all subgraphs of G which can be obtained by removing one edge are isomorphic to one of $G_1, \ldots, G_4$, it follows that G is critical.   □

There seems to be considerable similarity between graph 41 and graphs 22 and 24, even though the latter graphs are not bipartite. It seems probable that results like Lemma 8.2 could be proved for general classes of graphs which include graphs 22 or 24.

References

1. L. D. Andersen and A. J. W. Hilton, *Generalized latin rectangles*. II. *Embedding*, Discrete Math. **31** (1980), no. 3, 235–260.

2. L. D. Andersen, A. J. W. Hilton, and C. A. Rodger, *A solution to the embedding problem for partial idempotent latin squares*, J. London Math. Soc. (2) **26** (1982), no. 1, 21–27.

3. M. Behzad, *Graphs and their chromatic numbers*, Ph.D., Michigan State Univ., 1965.

4. M. Behzad, G. Chartrand, and J. K. Cooper, *The colour numbers of complete graphs*, J. London Math. Soc. (2) **42** (1967), 226–228.

5. C. Berge, *Graphs and hypergraphs*, 2nd revised ed., North-Holland Mathematical Library, vol. 6. North-Holland Publishing Co., Amsterdam-London; American Elsevier Publishing Co., Inc., New York, 1973

6. B. Bollobás, *Graph theory: An introductory course*, Springer-Verlag, New York, 1979.

7. _____, *Random graphs*, Academic Press, London, 1985.

8. V. Bryant and H. Perfect, *Independence theory in combinatorics. An introductory account with applications to graphs and transversals*, Chapman and Hall Math. Ser., Chapman & Hall, London-New York, 1980.

9. B. L. Chen and H. L. Fu, *The total colouring of graphs of order 2n and maximum degree 2n − 2*, Graphs Combin. **8** (1992), 119–123.

10. B. L. Chen, H. L. Fu, and H. P. Yap, *Total chromatic number of graphs of high degree*, manuscript.
11. A. G. Chetwynd and R. Häggkvist, *Some upper bounds on the total and list chromatic numbers of multigraphs*, J. Graph Theory **16** (1992) no. 5, 503–516.
12. A. G. Chetwynd and A. J. W. Hilton, *The chromatic class of graphs with at most four vertices of maximum degree*, Congr. Numer. **43** (1984) 221–248.
13. _____, *Regular graphs of high degree are 1-factorizable*, Proc. London Math. Soc. (3) **50** (1985) no. 2, 193–206.
14. _____, *The edge-chromatic class of regular graphs of degree 4 and their complements*, Discrete Appl. Math. **16** (1987) no. 2, 125–134.
15. _____, *Some refinements of the total chromatic number conjecture*, Congr. Numer. **66** (1988), 195–215.
16. _____, *1-factorizing regular graphs of high degree-an improved bound*, Discrete Math. **75** (1989) no. 1-3, 103–112.
17. _____, *The edge-chromatic class of graphs with large maximum degree, where the number of vertices of maximum degree is relatively small*, J. Combin. Theory Ser. B **48** (1990), 45–66.
18. A. G. Chetwynd, A. J. W. Hilton and Z. Zhao, *On the total chromatic number of graphs of high minimum degree*, J. London Math. Soc. (2) **44** (1991), 193–202.
19. M. M. Cropper, J. L. Goldwasser, and A. J. W. Hilton, *The scope of three colouring conjectures*, J. Combin. Math. Combin. Comput. (to appear).
20. P. Erdős and L. Pósa, *On the maximal number of disjoint circuits of a graph*, Publ. Math. Debrecen **9** (1962), 3–12.
21. S. Fiorini and R. J. Wilson, *Edge-colourings of graphs*, Res. Notes Math. **16**, Pitman, London; distributed by Fearon-Pitman Publishers, Inc., Belmont, Calif., 1977.
22. A. Hajnal and E. Szemerédi, *Proof of a conjecture of P. Erdős*, Combinatorial Theory and its Applications II (P. Erdős, A. Rényi and V. T. Sòs, eds.), Math. Soc. J. Bolyai, vol. 4, North-Holland, Amsterdam, 1970, pp. 601–623.
23. G. M. Hamilton and A. J. W. Hilton, *Graphs of maximum degree 3 and order at most 16 which are critical with respect to the total chromatic number*, J. Combin. Math. Combin. Comput. **10** (1991) 129–149.
24. A. J. W. Hilton, *A total chromatic number analogue of Plantholt's Theorem*, Discrete Math. **79** (1989/90), 169–175.
25. _____, *The total chromatic number of nearly complete bipartite graphs*, J. Combinatorial Theory Ser. B **52** (1991), 9–19.
26. _____, *Recent results on the total chromatic number*, Discrete Math. **111** (1993), 323–331.
27. A. J. W. Hilton and H. R. Hind, *The total chromatic number of graphs having large maximum degree*, Discrete Math. **117** (1993) no. 1-3, 127–140.
28. H. Hind, *An upper bound for the total chromatic number*, Graphs Combin. **6** (1990) no. 2, 153–159.
29. _____, *An upper bound for the total chromatic number of dense graphs*, J. Graph Theory **16** (1992) no. 3, 197–203.
30. A. V. Kostochka, *The total coloring of a multigraph with maximal degree 4*, Discrete Math. **17** (1977) no. 2, 161–163.
31. L. Lovász and M. D. Plummer, *Matching theory*, North-Holland Math. Stud. vol. 121; Annals of Discrete Mathematics, vol. 29, Budapest, 1986.
32. C. McDiarmid and B. Reed, *On total colourings of graphs*, J. Combin. Theory Ser. B **57**, (1993) no. 1, 122–203.
33. M. Molloy and B. Reed, *A bound on the total chromatic number*, Technical Report 304/96, Dept. Computer Science, Univ. Toronto.
34. T. Niessen and L. Volkman, *Class 1 conditions depending on the minimum degree and the number of vertices of maximum degree*, J. Graph Theory **14**, (1990), no. 2, 225–246.
35. A. Sánchez-Arroyo, *Determining the total colouring number is NP-hard*, Discrete Math. **78** (1989), no. 3, 315–319.
36. V. G. Vizing, *On an estimate of the chromatic class of a p-graph*, Diskret. Analiz **3** (1964), 25–30. (Russian)
37. _____, *Some unsolved problems in graph theory*, Uspekhi Mat. Nauk **23** (1968), no. 1-6, 117–134.

38. H. P. Yap, B. L. Chen and H. L. Fu, *Total chromatic number of graphs of order $2n + 1$ having maximum degree $2n - 1$*, J. London Math. Soc. **52** (1995) no. 3, 434–446.

DEPARTMENT OF ENGINEERING, READING UNIVERSITY, READING RG6 6AX, ENGLAND

DEPARTMENT OF MATHEMATICS, READING UNIVERSITY, READING RG6 6AX, ENGLAND
E-mail address: A.J.W.Hilton@reading.ac.uk

41, DAWSON CRESCENT, GEORGETOWN, ONTARIO, L7G 1H3, CANADA
E-mail address: h.r.hind@ieee.org

Centre de Recherches Mathématiques
CRM Proceedings and Lecture Notes
Volume **23**, 1999

Graphs Whose Radio Coloring Number Equals the Number of Nodes

Frank Harary and Michael Plantholt

For this note to be self-contained, we recall that a radio coloring of a graph G is a coloring of its n nodes in colors 1, 2, ... such that the colors of adjacent nodes differ by at least two, and any pair of nodes that are at distance 2 from each other in G are assigned colors that differ (by at least one). The radio coloring number of G, denoted $rc(G)$, is the minimum value of k for which there is a radio coloring of G using only colors in the set $\{1, 2, \ldots, k\}$. For graphs of order n, the radio coloring number can be as high as $2n - 1$ (for the complete graph) or as low as 1 (for its complement, the graph with no edges). We follow the graph terminology of [**2**].

Our purpose is to show that for almost all graphs, the radio coloring number is equal to the number of nodes, $rc(G) = n$. To do so, we make two observations in the form of lemmas; the first follows directly from the definition of radio coloring.

LEMMA 1. *If G has diameter at most 2, then $rc(G) \geq n$.*

LEMMA 2. *If $\overline{G}$, the complement of G, has a Hamilton path, then $rc(G) \leq n$.*

PROOF. Let v_1, v_2, ..., v_n be the nodes of a Hamilton path in $\overline{G}$. Because v_i is not adjacent to v_{i+1} in G for $i = 1$, ..., $n - 1$, we see that v_i and v_{i+1} may be assigned consecutive colors in a radio coloring of G. Thus the assignment of color h to node v_h gives a radio coloring. □

We also note the following two well-known asymptotic results, [**1, 5, 6**].

THEOREM A. *Almost all graphs have diameter 2.*

THEOREM B. *Almost all graphs are Hamiltonian.*

THEOREM 1. *For almost all graphs, the radio coloring number equals the number of nodes.*

PROOF. By Theorem B, the complements of almost all graphs are Hamiltonian. Thus we conclude from Theorems A and B that for almost all graphs, the diameter is two and the complement has a Hamilton path. The result then follows from Lemmas 1 and 2. □

1991 *Mathematics Subject Classification.* 05C15.
This is the final form of the paper.

Note that we have not answered the question originally asked in [3], that is, to characterize those graphs of order n for which $rc(G) = n$, and for which each color is used once in an optimal coloring. Two such graphs (both of diameter two) are the Petersen graph and the triangular prism. In fact Theorem 1 and its proof shows that this property ($rc(G) = n$ and each color is used once in an optimal coloring) holds for almost all diameter two graphs, and therefore for almost all graphs, so we would not expect that a reasonable characterization of these exists. However, there also exist graphs which have this property but which have diameter greater than 2, and it may be possible to characterize these graphs. One such example is constructed by beginning with a triangle with vertices v_1, v_2, and v_3, and two isolated vertices v_4 and v_5, and then adding edges $v_1 v_4$ and $v_2 v_5$. This is the self-complementary graph of the letter A, see [4].

We conclude with another observation, based on the correspondence between an n-coloring that uses each color once and a Hamilton path in the complement.

COROLLARY. *Let G be a graph with order n and diameter greater than 2. Suppose that $rc(G) = n$ and that each of the n colors is used once in an optimal coloring. Then:*

(i) *The complement $\overline{G}$ contains a Hamilton path, and*

(ii) *If the distance $d(x, y) > 2$ in G, then the graph G^* obtained by identifying nodes x and y has no Hamilton path in its complement.*

References

1. A. Blass and F. Harary, *Properties of almost all graphs and complexes*, J. Graph Theory **3** (1979), 225–240.
2. F. Harary, *Graph theory*, Addison-Wesley, Reading, MA, 1969.
3. _____, *Color costs of a graph and the radio coloring number*, private communication.
4. _____, *Typographs*, Visible Language **7** (1973), 199–208.
5. F. Harary and E. Palmer, *Graphical enumeration*, Academic Press, New York, 1973.
6. E. M. Palmer, *Graphical evolution: An introduction to the theory of random graphs*, Wiley, New York, 1985.

DEPARTMENT OF COMPUTER SCIENCE, NEW MEXICO STATE UNIVERSITY, LAS CRUCES, NM 88003, USA
   E-mail address: fnh@crl.nmsu.edu

DEPARTMENT OF MATHEMATICS, ILLINOIS STATE UNIVERSITY, NORMAL, IL 61790, USA
   E-mail address: mikep@mail.ilstu.edu

Centre de Recherches Mathématiques
CRM Proceedings and Lecture Notes
Volume **23**, 1999

The Height and Length of Colour Switching

Odile Marcotte and Pierre Hansen

ABSTRACT. Let G be a simple graph and C and D two proper colourings of G. The problem of colour switching consists of finding a sequence of vertex re-colourings that transforms C into D, all intermediate colourings being proper. The *height* of a colour switching is the number of additional colours used (i.e., colours that are neither in C nor in D). The *length* of a colour switching is the number of vertex recolourings in the switching. We study the minimum height and the minimum length of colour switchings in various classes of graphs. This problem is a simplified model of frequency reassignment in mobile telecommunication systems.

1. Introduction

Let $G = (V, E)$ be a loopless undirected graph. A *vertex-colouring* (or simply *colouring*) of G is a function from V to R, where R is a finite set of *colours*. The *cardinality* of this colouring is the cardinality of R itself. The *colour class i* (for some element i of R) is the set of vertices of G to which the colour i has been assigned. A colouring is said to be *proper* if for every edge uv of G, the colours assigned to u and v are different. The *chromatic number* of G is the smallest cardinality of a proper colouring of G. In the rest of this paper all colourings will be proper and we shall thus omit the qualifier "proper".

These definitions can be easily extended to the case of *weighted graphs*. A weighted graph G is a triple (V, E, w), where w is a weight function on the vertices. A colouring of the weighted graph G is an assignment of w_u distinct colours of R to each vertex u of G. A colouring is proper if for every edge uv of G, the w_u colours assigned to u and the w_v colours assigned to v are all different. The notions of colouring cardinality and chromatic number are defined in the same manner as for unweighted graphs.

Graph colouring has many applications in computer science and engineering, and in particular, the colouring problem for weighted graphs is a simplified version

1991 *Mathematics Subject Classification*. 05C50, 05C15, 94C15.

The work of the first author was supported by the Natural Sciences and Engineering Research Council of Canada (NSERC) under grant OGP 0009126.

The work of the second author was supported by ONR grant N00014-95-1-0917, NSERC grant OGP 0039682 and FCAR grant 32EQ 1048.

The authors would like to thank two anonymous referees for their helpful suggestions.

This is the final form of the paper.

of the problem of assigning frequencies to the cells of a mobile telecommunication system. In this model, the cells correspond to the vertices of a graph G, the reuse constraints to the edges of G, and the weight of vertex u (for each u) to the number of frequencies required in the corresponding cell. Thus there is a proper colouring of G with k colours if and only if there is an assignment of k frequencies to the cells that respects the reuse constraints. A large literature has been devoted to this problem, and we refer the reader to Hale [4] for an introduction to the problem of frequency assignment.

On the other hand, in a mobile telecommunication system, the assignment of frequencies to cells is rarely static because of the increase in communication demand. Therefore new frequencies have to be assigned to many cells, and even if the initial colouring was almost optimal, the new one (say, colouring C) might not be. It might thus be necessary to replace C by a better colouring, say D, and this requirement motivates our work and the following definitions. Given the (weighted or unweighted) graph G and a colouring C of G, we say that colouring D can be obtained from C by a *vertex recolouring* if for some vertex u, the sets of w_u colours assigned to u in C and D respectively have $w_u - 1$ colours in common, and if for any other vertex, the sets of colours assigned to this vertex in C and D are identical.

DEFINITION 1.1. Given two proper colourings C and D, the problem of *colour switching* consists of finding a sequence $(C = C_1, C_2, \ldots, C_p = D)$ in which C_i is a proper colouring of G and can be obtained from C_{i-1} (for $2 \leq i \leq p$) by a vertex recolouring. Such a sequence will be called a *switching* from C to D.

The problem of colour switching has so far received little attention in the scientific literature, with the exception of the paper by Barbéra and Jaumard [1]. The problem considered by Chee and Lim in [3] differs from ours in that the colouring D of the previous definition is only specified up to a permutation of colours. Their goal, then, is to find a permutation minimizing the number of vertices whose colour must be changed, and they show that it can be found in polynomial time by reducing the problem to a matching problem. In this paper we study colour switchings from the point of view of height and length. In the sequel S will denote the switching $(C = C_1, C_2, \ldots, C_p = D)$.

DEFINITION 1.2. The *height* of the switching $S = (C = C_1, C_2, \ldots, C_p = D)$ is the number of additional colours used by S (i.e., the number of colours used by S that are neither in C nor in D).

DEFINITION 1.3. The *length* of the switching $S = (C = C_1, C_2, \ldots, C_p = D)$ is the number of vertex recolourings in the switching (i.e., the number $p - 1$).

The paper is organized as follows. In Section 2 we give bounds on the height of switchings in unweighted and weighted graphs. In Section 3 we present results on the length of switchings of optimal or quasi-optimal height. In Section 4 we study recolourings of graphs that can be decomposed along a k-vertex cutset whose k vertices induce a clique, and we use this result to show that if G is a weighted k-tree and C and D are proper colourings of G, there exists a switching from C to D of height 1. Finally Section 5 contains a short conclusion.

2. Some Bounds on the Height of Switchings

In this section we consider two fixed colourings C and D of an undirected graph $G = (V, E)$. We may assume, without loss of generality, that the cardinality of C

(say, m) is at least that of D (say, n), that the colour classes of C and D are not empty and that any colour of D is also a colour of C. For if D includes colours that are not in C, we relabel colours of D that are not in C so that they are also in C and call the resulting colouring D'. We then observe that a switching from C to D can be obtained by concatenating any switching from C to D' and a switching of height 0 from D' to D. Note that while this procedure has no bearing on the height of the switching, the same cannot be said of its length. From now on we thus assume that the colours of C are labeled $1, 2, \ldots, m$ and those of D are labeled $1, 2, \ldots, n$. In the following definitions C and D denote two colourings of a graph G.

DEFINITION 2.1. An *entry* of the pair $\{C, D\}$ is a set of the form $U_{ij}(C, D) = \{v \in V \mid v \text{ has colour } i \text{ in } C \text{ and } j \text{ in } D\}$ for some $i \in \{1, 2, \ldots, m\}$ and $j \in \{1, 2, \ldots, n\}$. In other words, $U_{ij}(C, D)$ is the set of vertices of G whose initial colour is i and final colour j.

Some of the entries of the pair $\{C, D\}$ may be empty, and it turns out that the height of a switching from C to D depends upon the structure of the matrix formed by the entries.

DEFINITION 2.2. A *spike* of the pair $\{C, D\}$ is a set of entries of the form $U_{jj}(C, D), U_{j-1,j}(C, D), \ldots, U_{j-\ell,j}(C, D)$, where ℓ is a nonnegative integer, the entry $U_{j-\ell,j}(C, D)$ is not empty and every entry $U_{ij}(C, D)$ (for $1 \leq i < j - \ell$) is empty. The *spike index* of $\{C, D\}$ is the number of entries of the longest spike, i.e., the smallest ℓ for which all entries $U_{ij}(C, D)$ satisfying $i < j - \ell$ are empty.

To illustrate these definitions we introduce the graph $G_{m,n}^s$ defined as follows: the vertices of $G_{m,n}^s$ are couples of the form (i, j) satisfying $1 \leq i \leq m$, $1 \leq j \leq n$ and $i \geq j - s$, and its edges are the pairs $\{(i, j), (i', j')\}$ satisfying $i \neq i'$ and $j \neq j'$. There are two *standard* colourings of $G_{m,n}^s$: the first one, denoted C, assigns the colour i to vertex (i, j) (for all i and j), while the second one, denoted D, assigns the colour j to (i, j). By construction of $G_{m,n}^s$ and choice of C and D, the spike index of the pair $\{C, D\}$ is s. Note also that there is at most one vertex of $G_{m,n}^s$ in each entry of $\{C, D\}$.

Figure 1 contains a representation of $G_{9,9}^3$ in which entry (i, j) is located at the intersection of the ith row and jth column (row 1 is at the top of the figure and column 1 at the left). The colour of an entry is given by the number appearing in the table (i.e., the colour of entry $(2, 1)$ is 2 while that of $(3, 6)$ is 12). The colouring represented belongs to a switching from C to D and colours 10, 11 and 12 are additional colours.

Note that in general, the value of the spike index depends upon the labels of the colours in the colourings C and D. As we shall see below, it is desirable to relabel the colours in such a way that the spike index is as small as possible.

LEMMA 2.3. *Let us assume that C and D have identical colour classes. Then there exists a switching of height 1 from C to D.*

PROOF. Since C and D have identical colour classes, there exists a permutation π such that for each i ($1 \leq i \leq m$), the colour class i of C is identical to the colour class $\pi(i)$ of D. Because π is a product of transpositions, we can construct a switching from C to D by concatenating switchings from C_r to C_{r+1} (for $1 \leq r \leq q - 1$), where $C_1 = C$, $C_q = D$ and C_{r+1} is obtained from C_r by transposing two

1	10	10	10					
2	2	11	11	11				
3	3	3	12	12	12			
4	4	4	4	4	4	4		
5	5	5	5	5	5	5	5	
6	6	6	6	6	6	6	6	6
7	7	7	7	7	7	7	7	7
8	8	8	8	8	8	8	8	8
9	9	9	9	9	9	9	9	9

FIGURE 1. A representation of the graph $G_{9,9}^3$ and one of its colourings.

colours. To complete the proof, we observe that if C_{r+1} is obtained from C_r by transposing colours i and i', we can assign an additional colour (say, colour $m+1$) to all vertices in the colour class i' of C_r, then assign i' to all vertices in the colour class i and finally assign i to all vertices in the colour class $m+1$. This sequence of recolourings is actually a switching from C_r to C_{r+1} and uses one additional colour only (colour $m+1$). □

PROPOSITION 2.4. *Let us assume that the spike index of the pair $\{C, D\}$ is s. Then there exists a switching of height s from C to D.*

PROOF. Let us first assume that $2s$ is at most n. We begin by recolouring the vertices of the entries $U_{ij}(C, D)$ (for $1 \leq i \leq s$ and $i+1 \leq j \leq i+s$) with s additional colours. This can be done for instance by assigning colour $m+1$ to vertices of the entries $U_{1j}(C, D)$ for $2 \leq j \leq s+1$, colour $m+2$ to vertices of the entries $U_{2j}(C, D)$ for $3 \leq j \leq s+2$, and so on (see Fig. 1 for an illustration). For $1 \leq j \leq s$, we then assign the colour j to the jth *column* (i.e., the vertices of the entries $U_{ij}(C, D)$ for $1 \leq i \leq m$). Finally, we assign the colour $m+j-s$ to the jth column (for $s+1 \leq j \leq 2s$) and the colour $j-s$ to the jth column (for $2s+1 \leq j \leq n$). The process just described is indeed a switching because when we start assigning the colour r to the cells of the jth column, no other column contains a vertex having the colour r.

The colouring obtained by this process (called C') has the same colour classes as D. If s is equal to 0, C' is actually identical to D and the switching just described is of height 0. If s is greater than 0, we can recolour the jth column (for $s+1 \leq j \leq 2s$) by using colour $n-2s+j$, and the new colouring (called D') has the same colour classes and the same colours as D. By the previous lemma there is a switching of height 1 from D' to D. The concatenation of the switching of height s described in the previous paragraph, the switching of height 0 from C' to D' and the switching of height 1 from D' to D is the required switching of height s from C to D.

In the case where $2s$ is greater than n, the procedure is similar but somewhat simpler. We recolour the entries $U_{ij}(C, D)$ (for $1 \leq i \leq s$ and $i+1 \leq j \leq \min\{i+s, n\}$) with s additional colours, then assign the colour j to the jth column (for $1 \le j \le s$), the colour $m + j - s$ to the jth column (for $s + 1 \le j \le n - 1$) and finally the colour n to the nth column and the colour j to the jth column (for $s + 1 \le j \le n - 1$). □

Note that if D is an optimal colouring of G (i.e., n is actually the chromatic number of G), the previous proposition implies that for any colouring C of G, there exists a switching from C to D whose height is at most the chromatic number of G minus 1. Note also that for some pairs of colourings, the result of the previous proposition can be improved. For instance, if the chromatic number of the subgraph of G induced by $\{v \mid v \in U_{ij}(C, D)$ for some $i < j\}$ is equal to r, it is easy to check that the height of a switching from C to D is at most r. The following proposition shows that the above result is in a certain sense optimal.

PROPOSITION 2.5. *Let m, n and s be positive integers such that $m \ge n > s$. There exists a graph G and two colourings C and D of G such that*

    (a) *the cardinality of C is m,*
    (b) *the cardinality of D is n,*
    (c) *the spike index of the pair $\{C, D\}$ is s, and*
    (d) *the height of any switching from C to D is at least s.*

PROOF. Let $G_{m,n}^s$ be the graph introduced just after Definition 2.2, C the colouring of $G_{m,n}^s$ that assigns the colour i to vertex (i, j) and D the colouring that assigns the colour j to (i, j). Finally, let us fix a switching from C to D, and let t denote the smallest number of vertex recolourings after which exactly s columns contain two entries (i.e., two vertices) of the same colour. Let J denote the set of these columns. The construction of $G_{m,n}^s$ implies that at this point, there must be at least s distinct colours in the columns belonging to J, and none of these colours is assigned to any vertex (i, j) with $j \notin J$. Since J does not contain $s + 1$ columns, there is a column not belonging to J whose index (say, k) is at most $s + 1$. Because this column contains at least m entries (or vertices), all of which are assigned different colours, the number of distinct colours assigned to the vertices of $G_{m,n}^s$ after t vertex recolourings is at least $s + m$. This completes the proof of the proposition because at least s of these colours must be additional colours. □

To conclude this section, we observe that the above results can easily be extended to weighted graphs. For if $G = (V, E, w)$ is a weighted graph, we can construct an unweighted graph G' by replacing each vertex u of G by w_u vertices forming a clique in G'. There is an obvious one-to-one correspondence between colourings of G and colourings of G'. This correspondence induces a one-to-one correspondence between switchings and the framework outlined in this section can thus be applied to weighted graphs as well.

3. On the Length of Switchings of Optimal or Quasi-Optimal Height

The "matrix" interpretation of colour switching suggests that for each vertex v belonging to some $U_{ij}(C, D)$ $(i < j)$, either v or the vertices in $U_{ji}(C, D)$ must be assigned a temporary colour in any switching from C to D. Note that by "temporary colour" of v we mean a colour that is neither the colour of v in C (the original colour) nor its colour in D (the final colour). A temporary colour need not be additional. Let $U = \{v \mid v \in U_{ij}(C, D)$ for some $i < j\}$ and $U' = \{v \mid v \in U_{ij}(C, D)$ for some $i > j\}$. The above remark suggests that the minimal switching length is $|U'| + 2|U|$. Indeed, if s, the spike index of the pair $\{C, D\}$, is either 0 or $n - 1$, the switching described in the proof of Proposition 2.4 is of length $|U'| + 2|U|$. On the other hand, if $1 \leq s \leq n - 2$, it appears that we have to increase the height of the switching in order to achieve the same bound. The following lemma and proposition show this is indeed the case (under certain conditions) for switchings of the graph $G^s_{m,n}$ introduced in the previous section. To make the following arguments clearer, we shall say that a recolouring occurs at time t if it is recolouring nr. t in the switching in which it appears.

LEMMA 3.1. *Let C and D be the standard colourings of $G^s_{m,n}$, and U and U' be defined as above. The length of any switching from C to D is at least $|U'| + 2|U|$, where $|U| = s(n - (s + 1)/2)$ and $|U'| = n(m - (n + 1)/2)$.*

PROOF. Let $u_{ij} = (i, j)$ denote the only vertex of $G^s_{m,n}$ in U_{ij} (for all i and $j \leq i + s$), and consider the recolourings of u_{ij} and u_{ji} (for some $i < j \leq i + s$) in an arbitrary switching from C to D. By construction of $G^s_{m,n}$, these two vertices cannot be assigned the same colour, and we conclude that at least three recolourings are needed in order for u_{ij} and u_{ji} to be assigned their final colours. More precisely, if there is only one recolouring of u_{ij} in a given switching, and if this recolouring occurs at time t, there are at least two recolourings of u_{ji}: one occurring before t and assigning a temporary colour to u_{ji}, and another occurring after t and assigning its final colour to u_{ij}. A similar conclusion can be reached if there is only one recolouring of u_{ji}. Thus there are at least three recolourings for the pair $\{u_{ij}, u_{ji}\}$ and the length of any switching is at least $|U'| + 2|U|$ (i.e., the number of vertices in $|U|$ plus the number of vertices of $G^s_{m,n}$ that are not on the main diagonal). □

PROPOSITION 3.2. *Let m, n and s be positive integers such that $m \geq n \geq \max\{5, 2s + 2\}$. There exists a graph G and two colourings C and D of G such that*

(a) *the cardinality of C is m, that of D is n and the spike index of $\{C, D\}$ is s,*

(b) *there exists a switching from C to D of height $s + 1$ and length $|U'| + 2|U|$, but*

(c) *every switching from C to D of height s has length strictly greater than $|U'| + 2|U|$.*

PROOF. The graph having the properties just described is again $G^s_{m,n}$, and C and D are the standard colourings of $G^s_{m,n}$. Let $m + k$ (for $1 \leq k \leq s + 1$) denote the kth additional colour, and define the switching S as follows: for each $1 \leq j \leq n$, assign the colour $m + 1 + (j \mod (s + 1))$ to vertex $u_{j\ell}$ for $j + 1 \leq \ell \leq j + s$, and then assign colour j to vertices of the jth column that are not on the main diagonal (i.e., vertices of the form u_{ij} for $i \neq j$). It is straightforward to verify that S is indeed a switching of length $|U'| + 2|U|$.

Let now S be an arbitrary switching of length $|U'| + 2|U|$. We shall consider the colours assigned by S to vertices of G at time t for several values of t. We say that a column j is *fixed* at time t if it contains a vertex u_{ij} (for some $i \neq j$) that is assigned colour j at time t. Note that if a column j is fixed at time t, it is fixed for every $t' \geq t$ and vertices u_{ji} must be assigned a temporary colour or their final colour for every $t' \geq t - 1$.

Let also t_k (for $k = 1, 2, \ldots, s + 1, s + 2$) denote the smallest time at which k columns are fixed. Our observations imply that there are indices $\ell_1, \ell_2, \ldots, \ell_{s+1}$ such that at time t_k (for $k = 1, 2, \ldots, s + 1$), the set of columns that are fixed is precisely $\{\ell_1, \ell_2, \ldots, \ell_k\}$. If we let L denote $\{\ell_1, \ell_2, \ldots, \ell_{s+1}\}$, we can relabel the elements of L in order of increasing values, i.e., $L = \{i_1, i_2, \ldots, i_{s+1}\}$ and $i_k < i_{k+1}$ for $k = 1, 2, \ldots, s$. Finally let r be such that i_r is the same as l_{s+1}. Observe that i_1 must be equal to 1 since column $s + 2$ cannot be fixed before column 1. (This follows because vertex $u_{s+2,1}$ cannot be assigned a temporary colour in a switching of length $|U'| + 2|U|$.)

We now attempt to find vertices (i_k, j_k) that all have different (and temporary) colours at time $t_{s+1} - 1$ or $t_{s+2} - 1$. Observe that in this context, "temporary" colours are necessarily "additional" ones, because in a switching of length $|U'|+2|U|$, vertex (ℓ, ℓ) is never assigned a colour other than ℓ and thus vertex (i, j) (for $i, j \neq \ell$) is never assigned colour ℓ. We consider two cases.

CASE 1. $r \neq 1$

The assumption is equivalent to the statement $i_r \neq 1$. Let us define j_1 as

(3.1) $$\min\{j \mid j \notin (L\backslash\{i_r\}) \text{ and } i_1 < j \leq i_1 + s\}.$$

By assumption j_1 exists. For each $k = 2, \ldots, s + 1$ (starting with $k = 2$), we define j_k as

(3.2) $$\min\{j \mid j \notin (L\backslash\{i_r\}) \cup \{j_1, \ldots, j_{k-1}\} \text{ and } i_k < j \leq i_k + s\}.$$

Note that $L \cap \{i_k + 1, \ldots, i_k + s\}$ has at most $s + 1 - k$ elements because it does not include $i_1, i_2, \ldots, i_k$. We claim that j_k exists. For if i_r is at most i_k (i.e., $r \leq k$), j_1 is also at most i_k and $\{i_k + 1, \ldots, i_k + s\}\backslash\{i_{k+1}, \ldots, i_{s+1}, j_1, \ldots, j_{k-1}\}$ is nonempty. The same conclusion obtains if i_r is greater than $i_k + s$, or if i_r is greater than i_k and belongs to $\{j_1, j_2, \ldots, j_{k-1}\}$. Finally, if $i_k + 1 \leq i_r \leq i_k + s$ and $i_r \notin \{j_1, j_2, \ldots, j_{k-1}\}$, i_r can be chosen as the value of j_k.

To show the validity of the above procedure, we must also argue that each j_k is at most $2s + 2$. The claim follows easily if every $j \leq 2s + 1$ is of the form i_k or j_k. Let us assume this is not the case, and denote j the smallest index not of the form i_k or j_k (clearly $j \leq 2s + 1$). Then $i_p < j$ implies that $j_p < j$. Hence the number of columns of the form i_k for which $i_k < j$ is at most $\lceil (j-1)/2 \rceil$, and i_{s+1} is at least $s + 2 + \lfloor (j-1)/2 \rfloor$. But this implies in turn that i_{s+1} is equal to $s + 2 + \lfloor (j-1)/2 \rfloor$ and $\lceil (j-1)/2 \rceil = \lfloor (j-1)/2 \rfloor + 1$, since in order for i_{s+1} to be fixed, columns $1, 2, \ldots, i_{s+1} - s - 1$ must be fixed. We conclude that j_{s+1} and all the other j_k are at most $s + 2 + \lfloor (j-1)/2 \rfloor + s + 1 - \lceil (j-1)/2 \rceil = 2s + 2$.

Observe now that the vertices (i_k, j_k) (for $k = 1, 2, \ldots, s + 1$) must be assigned $s + 1$ distinct colours at all times since the indices $i_1, i_2, \ldots, i_{s+1}$, on one hand, and $j_1, j_2, \ldots, j_{s+1}$, on the other, are distinct. Note that at time $t_{s+1} - 1$, column i_k (for $k \neq r$) is fixed and column i_r is about to be fixed. Therefore (i_k, j_k) (for $k = 1, 2, \ldots, s + 1$) cannot have its original colour at time $t_{s+1} - 1$. Further, if $j_k \neq i_r$, (i_k, j_k) cannot have its final colour because j_k is not fixed by construction. If $j_k = i_r$ for some k, column j_k is not fixed at time $t_{s+1} - 1$ by definition of i_r, and (i_k, j_k) cannot have its final colour either. We conclude that the vertices (i_k, j_k) all have distinct temporary (and hence additional) colours.

CASE 2. $r = 1$

In this case, $i_r = 1$, i.e., column 1 is the last to be fixed in the first "group" of $s + 1$ fixed columns. Note that since entry $(s + 2, 1)$ can never have a temporary colour, column 1 must become fixed before column $s + 2$. Thus $i_{s+1} = s + 1$. We consider the colouring of G at time $t_{s+1} - 1$ (that is, just before column 1 becomes fixed). The vertices $(2, 1), (3, 1), \ldots, (s + 1, 1)$ and (i, j) (for $2 \leq i \leq s + 1$ and

1	2	3	4					
12	2	3	4	12				
11	2	3	4	11	11			
10	2	3	4	10	10	10		
5	2	3	4	5	5	5	5	
6	2	3	4	6	6	6	6	6
7	2	3	4	7	7	7	7	7
8	2	3	4	8	8	8	8	8
9	2	3	4	9	9	9	9	9

FIGURE 2. Another colouring of the graph $G_{9,9}^3$.

$s + 2 \leq j \leq s + i$) are assigned temporary colours at time $t_{s+1} - 1$ (see Fig. 2 for an illustration). If the number of temporary colours is greater than s, we are done. If there are only s temporary colours, each of these colours must be assigned to a *row*, i.e., a set of vertices of the form $(i, 1), (i, s+2), \ldots, (i, s+i)$, and the vertices $(i, s+2), \ldots, (i, s+i)$ will keep the colour they had at time $t_{s+1} - 1$ until time $t_{s+2} - 1$ (that is, until $s + 2$ columns have been fixed).

Let j denote the index of the $(s + 2)$th column to be fixed. j is comprised between $s + 2$ and $2s + 2$. If $j \leq 2s + 1$, vertices $(2, s+2), (3, s+3), \ldots, (s+1, 2s+1), (j, 2s+2)$ all have distinct and temporary colours at time $t_{s+2} - 1$. If $j = 2s + 2$, we note that vertices $(2, s+2), (3, s+2), \ldots, (s+1, s+2)$ have distinct and temporary colours at time $t_{s+2} - 1$, and that in order for column $2s + 2$ to become fixed at time t_{s+2}, vertex $(2s+2, 2s+1)$ (if $s > 1$) or $(2s+2, 2s+3)$ (if $s = 1$) must be assigned a $(s+1)$th temporary colour at time $t_{s+2} - 1$. In both cases, the switching S uses at least $s + 1$ additional colours. □

Various experiments with switchings will convince the reader that the hypothesis that $m \geq n \geq \max\{5, 2s + 2\}$ is quite unnecessary, but that if m and n are both equal to $s + 2$, only s temporary colours are needed to construct a switching. We are thus led to the following conjecture.

CONJECTURE 3.3. In the above proposition the hypothesis "$m \geq n \geq \max\{5, 2s + 2\}$" can be replaced by "$m \geq n \geq \max\{5, s + 2\}$ and m and n are not both equal to $s + 2$".

It is interesting to note that with one extra additional colour, one can achieve a very important reduction in the length of the switching (compare the lengths of the switchings described in Propositions 2.4 and 3.2).

4. Composition of Switchings of Bounded Height

It is well known (see for instance Berge [2]) that if a graph G has a k-vertex cutset that is also a clique, then the vertices of G can be coloured with the same number of colours as its "lobes" (the definition of lobe is given below). In this section we prove a similar result for switchings. Let $G = (V, E)$ be a graph and

K a nonempty subset of vertices of G that induces a clique. We say that G can be *decomposed* along K if V can be partitioned into nonempty subsets V_1, V_2 and K such that there is no edge in G with one endpoint in V_1 and the other in V_2. We denote by G_i (for $i = 1, 2$) the subgraph of G induced by $V_i \cup K$ and call G_1 and G_2 the *lobes* of G. Note that this decomposition might not be unique and the following results are independent of the chosen decomposition. We now show that the height of switchings for G is closely related to the heights of switchings for G_1 and G_2.

PROPOSITION 4.1. *Assume that the graph G can be decomposed into lobes G_1 and G_2 and C and D are two colourings of G. Assume also that there exists, for $i = 1, 2$, a switching S_i of height h_i from C_i to D_i (where C_i and D_i denote the restrictions of C and D, respectively, to G_i). Then there exists a switching from C to D of height $\max\{h_1, h_2, 1\}$.*

PROOF. Let K denote the clique along which G can be decomposed into G_1 and G_2. By Lemma 2.3 we may assume that any vertex of K has the same colour in D as in C. The colours assigned to vertices of G_2 in C will be called *original colours* of G_2. We shall construct a switching from C to C' and another from C' to D, where C' is the colouring of G whose restriction to G_1 is D_1 and restriction to G_2 is C_2.

The switching from C to C' is obtained from S_1 by inserting into it vertex recolourings of vertices of V_2, and it has the property that the colour classes of G_2 are the same in C' as in C. Also, immediately before recolouring a vertex of G_1, the colour classes of G_2 are the same as in C. To transform S_1, we consider two cases. If a vertex v of K is recoloured and assigned a colour (say colour i) that is not present in V_2, we insert into S_1, after this recolouring, recolourings that assign colour i to each vertex in the same colour class as v (colour class being understood with respect to colouring C_2).

If a vertex v of K is assigned a colour i that is present in V_2, we let j denote the current colour of v and use a suitable colour ℓ to transpose colour classes i and j in G_2. More precisely, we assign colour ℓ to vertices of V_2 that have colour i, assign colour i to v and each vertex in the same colour class as v (with respect to C_2) and finally assign colour j to each vertex of V_2 in colour class ℓ. If the current colours of G_2 are the original colours of G_2, we may choose as ℓ any additional colour. Note that if S_1 does not use any additional colour, colour ℓ is the unique additional colour used in the switching from C to C'. On the other hand, if at least one of the current colours is not original, we may choose as ℓ an original colour of G_2 that is not used in the current colouring of G_2.

At the end of the process just described, the vertices of K have their original colours and the switching from C' to D can be constructed in a similar fashion the previous one. This completes the proof of the proposition. $\square$

Note that the previous proposition can be trivially extended to weighted graphs, and that given colourings C and D of an arbitrary clique, there exists a switching from C to D of height 1. This suggests that we can use the proposition to derive the height of switchings in k-trees. Recall that a k-tree (where k is fixed) may be obtained from a clique of size k by successively adding to the current graph a vertex whose neighbours are the elements of a k-clique (see Rose [**5**] for another characterization of k-trees).

COROLLARY 4.2. *Let G be a weighted k-tree and C and D two colourings of G. There exists a switching from C to D of height 1.*

5. Conclusion

We have studied the problem of colour switching, which has so far received little attention in the literature. We have presented results on the height and length of switchings for a class of graphs that are "extremal" for this problem, and shown the relevance of the spike index for determining the minimal switching height. Many open questions need to be tackled. For instance, what can one say about the height and length of switchings in partial k-trees? In particular, it would be interesting to study the colour switchings of series-parallel graphs, bipartite graphs and line graphs. Finally, it will be necessary to explore the trade-off between height and length for various classes of graphs.

References

1. Vincent Barbéra and Brigitte Jaumard, *Design of an efficient channel block retuning*, Les Cahiers du GERAD, no. G-98-10 (submitted for publication).
2. Claude Berge, *Graphes et hypergraphes*, Dunod, Paris, 1970.
3. Yeow Meng Chee and Andrew Lim, *The algorithmic complexity of colour switching*, Inform. Process. Lett. **43** (1992), 63–68.
4. W. Hale, *Frequency assignment*, Proc. IEEE **68** (1980), 1497–1514.
5. D. J. Rose, *On simple characterization of k-trees*, Discrete Math. **7** (1974) 317–322.

GERAD AND DÉPARTEMENT D'INFORMATIQUE, UNIVERSITÉ DU QUÉBEC À MONTRÉAL, C.P. 8888, SUCCURSALE CENTRE-VILLE, MONTRÉAL (QC) H3C 3P8
    E-mail address: `odile@crt.umontreal.ca`

GERAD AND ÉCOLE DES HAUTES ÉTUDES COMMERCIALES, 3000 CHEMIN DE LA CÔTE-SAINTE-CATHERINE, MONTRÉAL (QC), H3T 2A7
    E-mail address: `pierreh@crt.umontreal.ca`

Centre de Recherches Mathématiques
CRM Proceedings and Lecture Notes
Volume **23**, 1999

Characteristic Polynomials in the Theory of Polyhedra

Horst Sachs

ABSTRACT. It is well known that, in certain areas of graph theory, the characteristic polynomial of a graph is a powerful tool—but, if properties of polyhedral graphs are concerned, then it often does not help. However, there is a way to define invariant polynomials especially for polyhedral (or plane) graphs. The starting point is Kasteleyn's classical theorem which allows the number of perfect matchings of a plane graph P to be calculated from the determinant of a somewhat modified adjacency matrix of P. By generalizing Kasteleyn's original concepts some polynomials for P are obtained which prove invariant under all relevant transformations. The problem is to find out what polyhedral properties are reflected by these invariants.

Applying these concepts to the carbon skeletons of polycyclic hydrocarbon molecules, we are led to a polynomial which essentially coincides with Hückel's characteristic polynomial based on molecular-orbital theory: this means that Hückel's polynomial can also be based on resonance theory, i.e., on Kekulé's much more elementary model.

1. Introduction

The graphs G to be considered are finite and have no loops or multiple edges; especially, if G is a directed graph then it does not have pairs of parallel or antiparallel arcs (directed edges). If not otherwise stated, G is assumed to be connected.

Notation.

$\mathbb{V}$: set of vertices of G.
$\mathbb{E}$: set of edges, or arcs, of G.
$\mathbb{F}$: set of faces of a plane graph G.
$v: = |\mathbb{V}|$.
$e: = |\mathbb{E}|$.
$f: = |\mathbb{F}|$.
z: cyclomatic number (dimension of the cycle space) of G. Recall that $z = e - v + 1$; for plane graphs, by Euler's formula, $z = f - 1$.

1991 *Mathematics Subject Classification.* Primary: 05C50; Secondary: 05C10, 05C70.
This is the final form of the paper.

$\mathbf{A} = (a_{ik})$: adjacency matrix of G:

$$a_{ik} = \begin{cases} 1 & \text{if there is an edge connecting vertices } i \text{ and } k, \\ & \text{or an arc going from } i \text{ to } k, \\ 0 & \text{otherwise.} \end{cases}$$

$\mathbf{D} = (d_{ik})$: degree matrix of an undirected graph G:

$$d_{ik} = \begin{cases} \text{degree (valency) of vertex } i \text{ if } i = k, \\ 0 \quad \text{otherwise.} \end{cases}$$

$\mathbf{L} = \mathbf{D} - \mathbf{A}$: Laplacian matrix of an undirected graph G.
$\mathbf{I}$: identity matrix.
$\mathbf{M}^{\top}$: transpose of matrix $\mathbf{M}$.
$\overline{\mathbf{M}}$: complex-conjugate of matrix $\mathbf{M}$.
spec $\mathbf{M}$: spectrum of matrix $\mathbf{M}$.
Pf($\mathbf{S}$): Pfaffian of a skew-symmetric matrix $\mathbf{S}$.
$f(\text{G}; x) = |x\mathbf{I} - \mathbf{A}|$: (ordinary) characteristic polynomial of G.
$l(\text{G}; x) = |x\mathbf{I} - \mathbf{L}|$: Laplacian polynomial of G.

It is a well-established fact that $f(\text{G}; x)$ and $l(\text{G}; x)$ are powerful tools in many fields of investigation (for a comprehensive updated overview, see [1]), but it is also known that there are problem fields in which these concepts are only of restricted value. In particular, in the theory of planar (or plane) graphs, only few applications are known: the definition of these polynomials is simply too general[1]. Therefore, the question may be raised whether it is possible to define similar polynomials that are more "tailor-made" for plane graphs, especially for polyhedra.

First let us define a somewhat modified adjacency matrix $\widetilde{\mathbf{A}} = \widetilde{\mathbf{A}}(\mathbf{G}) = (\tilde{a}_{ik})$ for a directed graph G which may be called the *skew adjacency matrix* of G:

$$\tilde{a}_{ik} = \begin{cases} 1 & \text{if there is an arc from } i \text{ to } k, \\ -1 & \text{if there is an arc from } k \text{ to } i, \\ 0 & \text{otherwise.} \end{cases}$$

Clearly, $\widetilde{\mathbf{A}}^{\top} = -\widetilde{\mathbf{A}}$ ($\widetilde{\mathbf{A}}$ is skew-symmetric) , thus

$$|\widetilde{\mathbf{A}}| = \begin{cases} 0 & \text{if } v \text{ is odd,} \\ \left(\text{Pf}(\widetilde{\mathbf{A}})\right)^2 & \text{if } v \text{ is even.} \end{cases}$$

Let us call

$$\tilde{f}(\text{G}; x) = |x\mathbf{I} - \widetilde{\mathbf{A}}|$$

the *skew (characteristic) polynomial* of G; its zeros, the *skew eigenvalues*, form the *skew spectrum* of G.

A square matrix $\mathbf{M}$ is called *skew-Hermitian* iff it satisfies $\mathbf{M}^{\top} = -\overline{\mathbf{M}}$. It can easily be proved that the nonzero eigenvalues of a skew-Hermitian matrix are pure imaginary. Hence

PROPOSITION 1. *The nonzero skew eigenvalues of a directed graph are pure imaginary.*

---

[1]Though, there are some important applications: let me just recall the methods of Pisanski and Fowler/Manolopoulos [3, 4, 6] for drawing the polyhedra of fullerenes (carbon cages) using the eigenvectors of $\mathbf{A}$; an interesting observation, also in connection with the study of fullerenes, has been made by P. W. Fowler (Exeter); I shall return to this topic at the end of this paper.

The coefficients of $\widetilde{f}(\mathrm{G};x)$ being real, the nonzero skew eigenvalues occur in pairs $x_j = iy_j$, $\bar{x}_j = -iy_j$ where the y_j are positive. This implies

PROPOSITION 2.

$$\widetilde{f}(\mathrm{G};x) = x^r \prod_{j=1}^{s}(x^2 + y_j^2) = x^r(x^{2s} + a_2 x^{2s-2} + \cdots + a_{2s}) \quad (r \geq 0, r+2s = v)$$

where the coefficients $a_{2\sigma}$ $(\sigma = 1, 2, \ldots, s)$ are positive.

For bipartite graphs, we can say a little bit more. Let G be a directed bipartite graph with bipartition $\mathbb{B}$, $\mathbb{W}$. Clearly, its skew adjacency matrix $\widetilde{\mathbf{A}}$ can be given the form

$$\begin{pmatrix} \mathbf{0} & \mathbf{U} \\ -\mathbf{U}^\top & \mathbf{0} \end{pmatrix}$$

where $\mathbf{U}$ has $b = |\mathbb{B}|$ rows and $w = |\mathbb{W}|$ columns; assume, w.l.o.g., $b \leq w$. Then

$$\widetilde{\mathbf{A}}^2 = \begin{pmatrix} -\mathbf{U}\mathbf{U}^\top & \mathbf{0} \\ \mathbf{0} & -\mathbf{U}^\top\mathbf{U} \end{pmatrix}$$

where $\mathbf{U}\mathbf{U}^\top$ and $\mathbf{U}^\top\mathbf{U}$ are symmetric square matrices of order b and w, respectively. Let the spectrum of $\widetilde{\mathbf{A}}$ be

$$\operatorname{spec}\widetilde{\mathbf{A}} = \{iy_1, -iy_1; \ldots; iy_s, -iy_s; 0, 0, \ldots, 0\}.$$

Then

$$\operatorname{spec}\widetilde{\mathbf{A}}^2 = \{-y_1^2, -y_1^2; \ldots; -y_s^2, -y_s^2; 0, 0, \ldots, 0\}$$
$$= \operatorname{spec}(-\mathbf{U}\mathbf{U}^\top) \cup \operatorname{spec}(-\mathbf{U}^\top\mathbf{U}).$$

As is well known, $\operatorname{spec}\mathbf{U}\mathbf{U}^\top$ and $\operatorname{spec}\mathbf{U}^\top\mathbf{U}$ are the same except for $w-b$ additional zeros in $\operatorname{spec}\mathbf{U}^\top\mathbf{U}$; therefore,

$$\operatorname{spec}\mathbf{U}\mathbf{U}^\top = -\operatorname{spec}(-\mathbf{U}\mathbf{U}^\top) = \{y_1^2, y_2^2, \ldots, y_s^2; 0, 0, \ldots, 0\}$$

(implying $s \leq b$) from which $\operatorname{spec}\widetilde{\mathbf{A}}$ can easily be retrieved.

For certain purposes, it may be advantageous to work with $\mathbf{U}\mathbf{U}^\top$ and its spectrum rather than with $\widetilde{\mathbf{A}}$ and its spectrum.

Define the *transpose* $\mathrm{G}^\top$ of a directed graph G to be the directed graph obtained from G by reversing all edge directions. Clearly, $\widetilde{\mathbf{A}}(\mathrm{G}^\top) = \widetilde{\mathbf{A}}^\top(\mathrm{G})$, hence

PROPOSITION 3. *The skew spectrum of a directed graph is invariant under transposition.*

Certain (images or embeddings of) undirected graphs have the property that their (combinatorial or geometrical) structure determines a (canonical) pair of distinguished reciprocal orientations, say $\pm\delta$, as, e.g., bipartite graphs B (if B has bipartition $\mathbb{B}$, $\mathbb{W}$ then the orientations are: from $\mathbb{B}$ to $\mathbb{W}$ and from $\mathbb{W}$ to $\mathbb{B}$) and graphs of a cyclic or "quasi-linear" structure as depicted in Fig. 1. Let, in these cases, G^δ denote one of the two distinguished directed versions of G. Then $\widetilde{f}(\mathrm{G}^\delta;x)$ is an invariant for the class of these graphs (or their geometric realizations).[2] I shall show that this idea can be extended so as to yield invariant polynomials for the class of all plane graphs where (topological) planeness (which is more discriminative than just planarity) is an essential hypothesis for defining these polynomials.

---

[2]It should, however, be noted that for bipartite graphs the resulting polynomial is essentially the same as the ordinary characteristic polynomial (see Section 10).

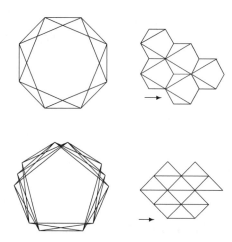

FIGURE 1

2. Switching an Orientation

Let G be an undirected graph, let ω be a possible orientation of its edges, and let G^ω denote the corresponding directed graph. Next we shall consider the set $\Omega = \Omega(G)$ of all possible orientations of G and study certain relations between them. Clearly, $|\Omega| = 2^e$.

To *switch* an orientation $\omega \in \Omega$ (or, briefly, to switch G^ω) with respect to some vertex V of G means to reverse the directions of all arcs of G^ω incident upon V. This operation will be called an *elementary switching*. A *switching* σ of G^ω is a sequence of elementary switchings; it is determined by a subset $\mathbb{X}$ of $\mathbb{V}$: the direction of an edge of G (i.e., of an arc of G^ω) is reversed by σ if and only if precisely one of its end vertices belongs to $\mathbb{X}$; therefore, let us write $\sigma = \sigma(\mathbb{X})$.

Note that every subset $\mathbb{X}$ of $\mathbb{V}$ determines a switching. Clearly, $\mathbb{X}$ and its complement $\mathbb{V} - \mathbb{X}$ determine the same switching: $\sigma(\mathbb{X}) = \sigma(\mathbb{V} - \mathbb{X})$. It is not difficult to prove that, for connected graphs,

$$\sigma(\mathbb{X}) = \sigma(\mathbb{Y}) \quad \text{if and only if either} \quad \mathbb{Y} = \mathbb{X} \quad \text{or} \quad \mathbb{Y} = \mathbb{V} - \mathbb{X}.$$

Thus the number of distinct orientations obtainable from a fixed orientation by means of switching is $\frac{1}{2}2^{|V|} = 2^{v-1}$.

Let us call two orientations *switching equivalent* iff one can be obtained from the other by switching. Each equivalence class containing precisely 2^{v-1} distinct orientations, the number of equivalence classes equals $2^e/2^{v-1} = 2^{e-v+1} = 2^z$.

The most important observation we make is that switching has no influence upon the polynomial $\widetilde{f}(G^\omega; x)$: indeed, switching with respect to vertex k is equivalent to multiplying both the k-th row and the k-th column of $\widetilde{\mathbf{A}}$ by -1. Thus $\widetilde{f}(G^\omega; x)$ is a class invariant with respect to switching.

By Proposition 4, transposition, too, does not change $\widetilde{f}(G^\omega; x)$. If G is bipartite with bipartition $\mathbb{B}$, $\mathbb{W}$ then, clearly, G^ω is transformed into its transpose by switching ω with respect to $\mathbb{B}$ or $\mathbb{W}$. Note that an odd circuit with directed edges cannot be transformed into its transpose by switching. This implies that transposition results from some switching operation if and only if G is bipartite.

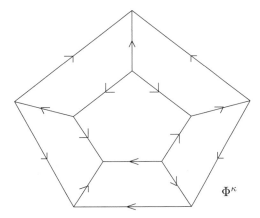

FIGURE 2. The graph Φ of the 5-prism equipped with a Kasteleyn orientation κ. We have
$$\widetilde{f}(\Phi^\kappa; x) = (x^2 + 1)(x^4 + 7x^2 + 11)^2,$$
thus, by Kasteleyn's theorem, the number m of perfect matchings (Kekulé patterns) of Φ is $m(\Phi) = \sqrt{\widetilde{f}(\Phi^\kappa; 0)} = 11$.

Thus if, in addition to switching, transposition is considered a feasible operation, then, for a nonbipartite graph G, the number of equivalence classes reduces to $\frac{1}{2}2^z = 2^{z-1}$.

3. Kasteleyn's Theorem

From now on, all graphs considered are assumed to be plane graphs (i.e., embeddings of planar graphs in the topological 2-sphere). In the plane, each edge has two *banks* , each arc has a left bank and a right bank. Let F be a face of a plane graph G. The *boundary length* $l(F)$ is the number of banks of G that belong to F. If G is equipped with an orientation ω, let $\lambda_\omega(F)$ denote the number of arcs of G^ω whose left bank belongs to F. Orientation ω is called *Kasteleyn with respect to F* iff $\lambda_\omega(F)$ is odd; it is called a *Kasteleyn orientation of* G iff it is Kasteleyn with respect to all faces of G (see Fig. 2). A famous theorem due to P. W. Kasteleyn [5] states that

(i) a plane graph G has a Kasteleyn orientation if and only if its number of vertices is even;

(ii) the number m of perfect matchings (linear factors) of a plane graph G with an even number of vertices is equal to

$$\mathrm{abs}\big(\mathrm{Pf}\big(\widetilde{\mathbf{A}}(G^\kappa)\big)\big) = \sqrt{\big|\widetilde{\mathbf{A}}(G^\kappa)\big|}$$

where κ is any Kasteleyn orientation of G.

Note that $\big|\widetilde{\mathbf{A}}(G^\kappa)\big| = \widetilde{f}(G^\kappa; 0)$.

In what follows this theorem will be put into a more general context.

A concept is called *chiral* iff it is not invariant under transposition, i.e., iff it does not allow "left" and "right" to be interchanged. Note that the concept of a Kasteleyn orientation is chiral whereas $\widetilde{f}(G^\omega; x)$ and $\big|\widetilde{\mathbf{A}}(G^\omega)\big|$ are not.

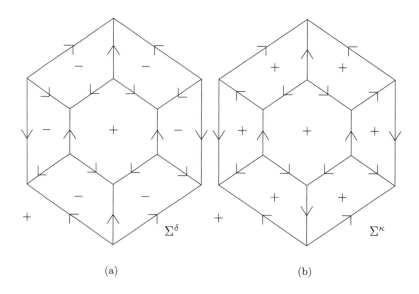

FIGURE 3. The graph Σ of the 6-prism equipped with orientations δ and κ which induce distinct character schemes. We have the equations

$$\widetilde{f}(\Sigma^\delta; x) = (x^2 + 9)(x^2 + 4)^2(x^2 + 1)x^4,$$

$$\widetilde{f}(\Sigma^\kappa; x) = (x^2 + 5)^2(x^2 + 2)^4$$

the latter of which implies $m(\Sigma) = \sqrt{5^2 \cdot 2^4} = 20$.

4. Character Schemes of Plane Graphs

Now consider any plane graph G equipped with an orientation ω. As there is no danger of confusion, the restriction of ω to some subgraph of G will briefly be also denoted by ω. By ω, the faces of G are partitioned into two (possibly, empty) sets:

$$\mathbb{F}_\omega^+ = \big\{ F \mid F \in \mathbb{F} \ \& \ \lambda_\omega(F) \text{ is odd } \big\}, \quad \mathbb{F}_\omega^- = \big\{ F \mid F \in \mathbb{F} \ \& \ \lambda_\omega(F) \text{ is even} \big\}.$$

Thus ω induces a *character scheme* χ_ω on the set of faces of G:

$$\chi_\omega(F) = \begin{cases} 1 & \text{iff } F \in \mathbb{F}_\omega^+, \\ -1 & \text{iff } F \in \mathbb{F}_\omega^-. \end{cases}$$

(see Fig. 3).

If G is not a tree then it has an edge E that lies on a circuit thus separating two faces, F' and F'', say. The omission of E in G, and of the corresponding arc in G^ω, makes F' and F'' coalesce into a new face, F^* say; clearly, $\lambda_\omega(F^*) = \lambda_\omega(F') + \lambda_\omega(F'') - 1$ implying

(1) $$\chi_\omega(F^*) = \chi_\omega(F') \cdot \chi_\omega(F'').$$

The edge omission process. Let T be any spanning tree of G. Successively (in an arbitrary order) omitting those edges from G that do not lie on T, we create a sequence of $z + 1$ spanning subgraphs of G, starting with G and ending with T.

Let $\widehat{G}$ be one of these subgraphs, let $\widetilde{F}$ be one of its faces, and let $\mathbb{F}^*(\widetilde{F})$ denote the set of those faces of G that are contained in $\widetilde{F}$: then, by extending (1),

$$(2) \qquad \chi_\omega(\widetilde{F}) = \prod_{F \in \mathbb{F}^*(\widetilde{F})} \chi_\omega(F);$$

in particular, if F_T denotes the only face of T,

$$(3) \qquad \chi_\omega(F_\mathrm{T}) = \prod_{F \in \mathbb{F}} \chi_\omega(F).$$

As each arc of T^ω contributes precisely one left bank to F_T, we have $\lambda_\omega(F_\mathrm{T}) = v - 1$, thus

$$(4) \qquad \chi_\omega(F_\mathrm{T}) = (-1)^v.$$

Equations (3) and (4) imply

LEMMA 1. $\prod_{F \in \mathbb{F}} \chi_\omega(F) = (-1)^v$ for any orientation ω of G.

5. Main Theorem for Character Schemes

Let T be an arbitrary spanning tree of G.

LEMMA 2. An orientation ω of G with given character scheme χ is uniquely determined by its restriction to T.

PROOF. Assume that orientations ω_1, ω_2 of G with $\chi_{\omega_1} = \chi_{\omega_2} = \chi$ coincide on T. Consider any edge E of G that does not lie on T and form the subgraph $\mathrm{G}(\mathrm{T}, E)$ of G consisting of T and the edge E; let $\mathrm{G}(\mathrm{T}, E)$ have faces F', F''. Assuming that E is the last edge omitted in the edge omission process described in Section 4, (2) applies to F', therefore,

$$\chi_{\omega_1}(F') = \chi_{\omega_2}(F') = \prod_{F \in \mathbb{F}^*(F')} \chi(F).$$

This implies that ω_1 and ω_2 coincide on E, thus $\omega_1 = \omega_2$. $\qquad \square$

THEOREM 1. Let φ be a function defined on $\mathbb{F}$ such that $\varphi(F) \in \{1, -1\}$.

(A) If φ is the character scheme for some orientation ω of G then φ satisfies the condition

$$(5) \qquad \prod_{F \in \mathbb{F}} \varphi(F) = (-1)^v.$$

(B) If φ satisfies (5) then it is the character scheme for precisely 2^{v-1} distinct orientations ω of G.

Proposition (A) is an immediate consequence of Lemma 1.

PROOF OF (B). We reverse the edge omission process described in Section 4 and start with an arbitrary spanning tree T of G; set $\mathrm{G}_0 = \mathrm{T}$. We select one of the 2^{v-1} distinct orientations of G_0, ω_0, say. Then, trivially, χ_{ω_0} with

$$\chi_{\omega_0}(F_\mathrm{T}) = (-1)^v = \prod_{F \in \mathbb{F}} \varphi(F) = \prod_{F \in \mathbb{F}^*(F_\mathrm{T})} \varphi(F)$$

is the character scheme of G_0 for the orientation ω_0. Adding successively the edges of G that are not in T (in an arbitrary order), we create a sequence $\mathrm{G}_0 = \mathrm{T}$, G_1, $\mathrm{G}_2, \ldots, \mathrm{G}_z = \mathrm{G}$. Assume that, for some i $(0 \le i < z)$, we have already constructed orientations $\omega_0, \omega_1, \ldots, \omega_i$ for $G_0, G_1, \ldots, G_i$, respectively, such that ω_{j-1} is the restriction of ω_j to G_{j-1} $(j = 1, 2, \ldots, i)$ and, for any face $\widetilde{F}$ of G_j,

$$\chi_{\omega_j}(\widetilde{F}) = \prod_{F \in \mathbb{F}^*(\widetilde{F})} \varphi(F) \quad (j = 0, 1, \ldots, i).$$

Now we add the edge E that is an edge of G_{i+1}, but not of G_i, to G_i. Edge E subdivides some face of G_i into two faces, F^1 and F^2 say, of G_{i+1}. It is easy to check that E can be directed in exactly one way such that the resulting orientation ω_{i+1} of G_{i+1} satisfies

$$\chi_{\omega_{i+1}}(F^\lambda) = \prod_{F \in \mathbb{F}^*(F^\lambda)} \varphi(F), \quad \lambda = 1, 2;$$

with this direction of E,

$$\chi_{\omega_{i+1}}(\widetilde{F}) = \prod_{F \in \mathbb{F}^*(\widetilde{F})} \varphi(F)$$

for all faces $\widetilde{F}$ of G_{i+1}.

   After z steps the procedure stops with an orientation $\omega = \omega_z$ satisfying $\chi_\omega(F) = \varphi(F)$ for all faces F of G. By Lemma 10, ω is uniquely determined by φ and ω_0.

   Starting with the 2^{v-1} distinct orientations ω_0 of T, we obtain 2^{v-1} (and no more) distinct orientations ω of G all satisfying $\chi_\omega = \varphi$.   □

6. Similarity of Orientations of a Plane Graph

   Call two orientations of G *similar* iff they induce the same character scheme. Clearly, similarity is an equivalence relation. By Theorem 1, all similarity classes have the same cardinality, namely, 2^{v-1}. Therefore, the number of similarity classes equals $2^e/2^{v-1} = 2^{e-v+1} = 2^z = 2^{f-1}$ (this is the number of functions φ satisfying condition (5)).

   Note that "vertex" and "face" and their numbers, v and f, are dual concepts, thus the last observation says that the number of similarity classes of a plane graph equals the cardinality of the similarity classes of its dual (and conversely). It may be an interesting task to investigate the relations between the sets of orientations of a plane graph and its dual.

   It is easy to see that a character scheme χ_ω of a plane graph does not change if an elementary switching is applied to ω, thus the character schemes are invariant under switching. Consequently, orientations that are switching equivalent are also similar. The similarity classes and the switching equivalence classes having the same cardinality (namely, 2^{v-1}), the converse is also true: similarity is just switching equivalence restricted to the class of plane graphs. It follows that the polynomial $\widetilde{f}(G^\omega; x)$ is a class invariant of similarity. Therefore, for any character scheme χ and any orientation ω such that $\chi_\omega = \chi$, let us denote the polynomial $\widetilde{f}(G^\omega; x)$ by $\widetilde{f}_\chi(G; x)$.

7. Plane Graphs with a Canonical Bipartition of Their Face Set

   Depending on the class of plane graphs to be investigated and on the character of the problems to be pursued, the set $\mathbb{F}$ of faces may have a distinguished (or: canonical) bipartition $\mathbb{F}_1, \mathbb{F}_2$ where one of these sets may be empty. In the theory of polyhedra and in the chemistry of polycyclic compounds (especially, hydrocarbons

TABLE 1

class of plane graphs	$\mathbb{F}_1$	$\mathbb{F}_2$
polyhedra	odd faces	even faces
bipartite polyhedra	$(4n+2)$-gons	$4n$-gons
Eulerian polyhedra[a]	black faces	white faces
(p,q)-polyhedra[b]	p-gons	q-gons
graphs of (nearly) plane molecules	the finite faces	the infinite face
graphs of benzenoids[c] and general polycyclic alternant systems[d] and their subgraphs	$l(F) \equiv 2, \mod 4$	$l(F) \equiv 0, \mod 4$
graphs of carbon cages (fullerenes)[e]	pentagons	hexagons
graphs of tubulenes[f] (cylindrical hexagonal systems)	hexagons	the two "holes" at the ends of the tubule
...	...	...

[a]$\mathbb{F}$ has a unique bipartition (into the classes of "black" and of "white" faces, say) such that no two faces of the same class have a boundary edge in common.
[b]simple polyhedra having only p-gons and q-gons as their faces.
[c]benzenoids are polycyclic hydrocarbons, their carbon skeletons are hexagonal systems, see Fig. 4.
[d]see Figs. 3a, 5.
[e]a subset of the set of $(5,6)$-polyhedra.
[f]see [**7**].

and carbon cages) there are many examples of such a situation some of which are listed in Table 1.

Let G be an undirected plane graph and let φ be a function defined on $\mathbb{F}$ such that

$$\varphi(F) = \begin{cases} \varepsilon_1 & \text{if } F \in \mathbb{F}_1, \\ \varepsilon_2 & \text{if } F \in \mathbb{F}_2, \end{cases}$$

where $\varepsilon_1, \varepsilon_2 \in \{+1, -1\}$.

If φ satisfies (5) then it is a character scheme giving rise to the skew characteristic polynomial $\widetilde{f}_\varphi(G; x)$ for the class of graphs G under consideration. The possible choices of the pair $(\varepsilon_1, \varepsilon_2)$ are given in Table 2 where $f_i = |\mathbb{F}_i|$, $i = 1, 2$ ("+ −" stands for "$(+1, -1)$", etc.).

Under a transposition T, the character of an odd face is changed whereas the character of an even face stays the same. Now Proposition 3 yields

PROPOSITION 4. *If the character schemes* χ, χ' *satisfy*

$$\chi'(F) = \begin{cases} \chi(F) & \text{if } F \text{ is even}, \\ -\chi(F) & \text{if } F \text{ is odd}, \end{cases}$$

then $\widetilde{f}_{\chi'}(G; x) = \widetilde{f}_\chi(G; x)$.

TABLE 2

f_1 \ f_2	v even		v odd	
	even	odd	even	odd
even	+ + + − − + − −	+ + + −		− + − −
odd	+ + − +	+ + − −	+ − − −	+ − − +

TABLE 3

f_2 \ v	even	odd
even	+ + − −	
odd	+ +	− −

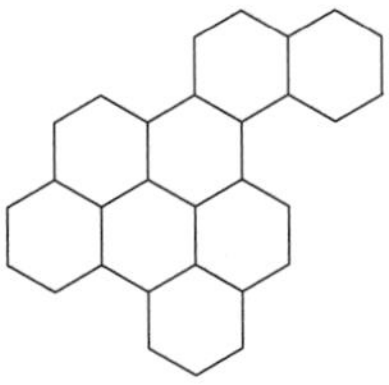

FIGURE 4

Let, in particular, $\mathbb{F}_1$ and $\mathbb{F}_2$ be the set of odd faces and the set of even faces, respectively. Note that $f_1 = |\mathbb{F}_1|$ is, necessarily, even. By Proposition 4, we may choose ε_1 so that $\varepsilon_1 = \varepsilon_2$. The remaining possible choices of $(\varepsilon_1, \varepsilon_2)$ are given in Table 3.

Note that in fullerene polyhedra, all vertices have valency 3, thus v is even.

In the theory of hydrocarbons, the graphs of polycyclic alternant (i.e., bipartite) systems (Figs. 4, 5) are of some interest. The distinction between $(4n + 2)$-gons and $4n$-gons being of particular relevance, let $\mathbb{F}_1$ and $\mathbb{F}_2$ be the set of $(4n + 2)$-gons and of $4n$-gons, respectively; set $f_i = |\mathbb{F}_i|$ $(i = 1, 2)$. It can easily be proved that $v \equiv f_2 \mod 2$. Table 4 lists the possible choices of $(\varepsilon_1, \varepsilon_2)$.

Note that Table 4 extends to the class of all connected bipartite plane graphs G; note also that there is precisely one choice for the characters ε_1, ε_2 that is always possible, namely, $(\varepsilon_1, \varepsilon_2) = (+1, -1)$. Therefore, we shall call the corresponding character scheme $\chi^d(G)$ with

$$(6) \qquad \chi^d(G; F) = \begin{cases} 1 & \text{if } l(F) \equiv 2, \mod 4, \\ -1 & \text{if } l(F) \equiv 0, \mod 4, \end{cases}$$

the *distinguished* character scheme of G (see Fig. 3a).

We shall return to this topic in Section 10.

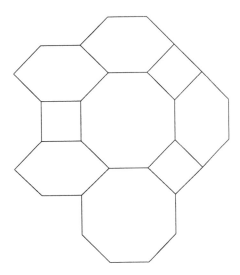

FIGURE 5

TABLE 4

f_2 \\ f_1	even	odd
even	+ + + − − + − −	+ + + −
odd	+ − − −	+ − − +

Note that Table 4 extends to the class of all connected bipartite plane graphs G; note also that there is precisely one choice for the characters ε_1, ε_2 that is always possible, namely, $(\varepsilon_1, \varepsilon_2) = (+1, -1)$. Therefore, we shall call the corresponding character scheme $\chi^d(G)$ with

(6)
$$\chi^d(G; F) = \begin{cases} 1 & \text{if } l(F) \equiv 2, \quad \text{mod } 4, \\ -1 & \text{if } l(F) \equiv 0, \quad \text{mod } 4, \end{cases}$$

the *distinguished* character scheme of G (see Fig. 3a).

We shall return to this topic in Section 19.

15. The Character Schemes of the Connected Subgraphs

Let G be a plane undirected graph, $\omega \in \Omega(G)$ and $\chi_\omega = \chi$. According to Proposition 3, the corresponding skew characteristic polynomial is

$$\widetilde{f}_\chi(G; x) = \widetilde{f}(G^\omega; x) = x^r(x^{2s} + a_2 x^{2s-2} + \cdots + a_{2s}).$$

By a classical theorem, the coefficient $a_{2\sigma}$ equals the sum of all principal minors of order 2σ of $\widetilde{\mathbf{A}}(G^\omega)$. These minors are the determinants of the $\binom{v}{2\sigma}$ skew adjacency matrices of the induced directed subgraphs on 2σ vertices of G^ω. If H^ω is such a *and any face $\widetilde{F}$ of $\widehat{G}$,*

$$\chi_\omega(\widetilde{F}) = (-1)^{v(\widetilde{F})} \prod_{F \in \mathbb{F}^*(\widetilde{F})} \chi_\omega(F).$$

COROLLARY 1. *Similar orientations of* G *induce similar orientations on each connected subgraph* $\widehat{G}$ *of* G.

Therefore, we may briefly say that a character scheme χ of G *induces* a character scheme $\widehat{\chi}$ on $\widehat{G}$.

PROOF OF THEOREM 2. Arrange those faces $\widetilde{F}$ of $\widehat{G}$ that contain at least one edge of G in their interior as a sequence $\widetilde{F}_1, \widetilde{F}_2, \ldots, \widetilde{F}_m$. If $m = 0$ then there is nothing to prove; assume $m \geq 1$. Form the sequence of subgraphs $G^0 = G$, G^1, $G^2, \ldots, G^m = \widehat{G}$ such that G^μ is a subgraph of $G^{\mu-1}$ and G^μ has exactly one face—namely, $\widetilde{F}_\mu$—that contains at least one edge of $G^{\mu-1}$ in its interior ($\mu = 1, 2, \ldots, m$). (Each face of G^μ, different from $\widetilde{F}_\mu$, is also a face of $G^{\mu-1}$; each face of $G^{\mu-1}$ that is not a face of G^μ is contained in $\widetilde{F}_\mu$.)

Let χ^1_ω be the character scheme of G^1 induced by ω. Set $\overline{\mathbb{F}}^*(\widetilde{F}_1) = \mathbb{F} - \mathbb{F}^*(\widetilde{F}_1)$. Note that $\mathbb{F} = \mathbb{F}^*(\widetilde{F}_1) \cup \overline{\mathbb{F}}^*(\widetilde{F}_1)$, $\mathbb{F}^1 = \{\widetilde{F}_1\} \cup \overline{\mathbb{F}}^*(\widetilde{F}_1)$ where $\mathbb{F}^1$ is the set of faces of G^1. For any face $\widetilde{F} \in \overline{\mathbb{F}}^*(\widetilde{F}_1)$, trivially,

$$v(\widetilde{F}) = 0, \quad \mathbb{F}^*(\widetilde{F}) = \{\widetilde{F}\}, \quad \chi^1_\omega(\widetilde{F}) = \chi_\omega(\widetilde{F}) = (-1)^{v(\widetilde{F})} \prod_{F \in \mathbb{F}^*(\widetilde{F})} \chi_\omega(F).$$

Further, as a consequence of Lemma 1 if applied to G and G^1,

$$\prod_{F \in \mathbb{F}} \chi_\omega(F) = \prod_{F \in \mathbb{F}^*(\widetilde{F}_1)} \chi_\omega(F) \cdot \prod_{F \in \overline{\mathbb{F}}^*(\widetilde{F}_1)} \chi_\omega(F) = (-1)^v,$$

$$\prod_{F \in \mathbb{F}^1} \chi^1_\omega(F) = \chi^1_\omega(\widetilde{F}_1) \cdot \prod_{F \in \overline{\mathbb{F}}^*(\widetilde{F}_1)} \chi_\omega(F) = (-1)^{v - v(\widetilde{F}_1)},$$

implying $\chi^1_\omega(\widetilde{F}_1) = (-1)^{v(\widetilde{F}_1)} \prod_{F \in \mathbb{F}^*(\widetilde{F}_1)} \chi_\omega(F)$.

This proves the assertion for the subgraph G^1 of G. Iterating this way of reasoning, we eventually arrive at the proof of Theorem 2.   □

9. The Kasteleyn Polynomial

The probably most important skew characteristic polynomial arises if all characters are taken to be 1 (briefly: $\chi = \mathbf{1}$) (Figs. 2 and 3b). By Theorem 1, this is possible if and only if v is even. This condition is not as restrictive as it may seem to be: graphs of simple polyhedra (which are regular of degree 3), graphs that have a perfect matching, and many other graphs of particular interest all have an even number of vertices. The similarity class corresponding to $\chi = \mathbf{1}$ is the class of Kasteleyn orientations: therefore, let us call $\chi = \mathbf{1}$ the *Kasteleyn character scheme* and denote the polynomial $\widetilde{f}_\mathbf{1}(G; x)$ by $K(G; x)$:

$$K(G; x) = \widetilde{f}_\mathbf{1}(G; x) = x^v + k_2 x^{v-2} + \cdots + k_v.$$

According to Kasteleyn's theorem, $\sqrt{k_v}$ equals the number of perfect matchings of G, but $K(G; x)$ contains much more information about the (matching) structure of G than just its last coefficient: therefore, $K(G; x)$ deserves to be thoroughly studied.

TABLE 5

G	$f(G; x)$	$l(G; x)$	$K(G; x)$
4-path	$x^4 - 3x^2 + 1$	$x(x - 2) \times$ $(x^2 - 4x + 2)$	$x^4 + 3x^2 + 1$
3-claw	$x^2(x^2 - 3)$	$x(x - 1)^2(x - 4)$	$x^2(x^2 + 3)$
square	$(x + 2)x^2(x - 2)$	$x(x - 2)^2(x - 4)$	$(x^2 + 2)^2$
hexagon	$x^6 - 6x^4 + 9x^2 - 4 =$ $(x^2 - 1)^2(x^2 - 4)$	$x(x - 1)^2 \times$ $(x - 3)^2(x - 4)$	$x^6 + 6x^4 + 9x^2 + 4 =$ $(x^2 + 1)^2(x^2 + 4)$

The platonic solids

tetrahedron	$(x + 1)^3(x - 3)$	$x(x - 4)^3$	$(x^2 + 3)^2$
octahedron	$(x + 2)^2 x^3(x - 4)$	$x(x - 4)^3(x - 6)^2$	$(x^2 + 4)^3$
hexahedron (cube)	$(x + 3)(x + 1)^3 \times$ $(x - 1)^3(x - 3)$	$x(x - 2)^3 \times$ $(x - 4)^3(x - 6)$	$(x^2 + 3)^4$
icosahedron	$(x + 1)^5(x - 5) \times$ $(x^2 - 5)^3$	$x(x - 6)^5 \times$ $(x^2 - 10x + 20)^3$	$(x^2 + 5)^6$
dodecahedron	$(x + 2)^4 \times$ $x^4(x - 1)^5 \times$ $(x - 3)(x^2 - 5)^3$	$x(x - 2)^5 \times$ $(x - 3)^4(x - 5)^4 \times$ $(x^2 - 6x + 4)^3 \times$ $(x^2 + 1)^6$	$(x^2 + 6)^4(x^2 + 1)^6$

The football graph F

F is the graph fullerene C_{60}	$(x + 2)^4(x - 1)^9 \times$ $(x - 3) \times$ $(x^2 + 3x + 1)^3 \times$ $(x^2 + x - 1)^5 \times$ $(x^2 + x - 4)^4 \times$ $(x^2 - x - 3)^5(x^4 - $ $3x^3 - 2x^2 + 7x + 1)^3$	$x(x - 2)^9 \times$ $(x - 5)^4 \times$ $(x^2 - 5x + 3)^5 \times$ $(x^2 - 7x + 11)^5 \times$ $(x^2 - 7x + 8)^4 \times$ $(x^2 - 9x + 19)^3 \times$ $(x^4 - 9x^3 + 25x^2 - $ $22x + 4)^3$	$(x^2 + 5)^6 \times$ $(x^4 + 9x^2 + 10)^4 \times$ $(x^4 + 3x^2 + 1)^8$

The last entry entails $m(F) = 5^3 \cdot 10^2 = 12500$.

In Table 5, some simple examples are listed.

Note that, for even v, the character schemes form a group under elementwise multiplication, with the Kasteleyn character scheme acting as the unity element. Is there any nontrivial relation between $\widetilde{f}_{\chi'}(G; x)$, $\widetilde{f}_{\chi''}(G; x)$ and $\widetilde{f}_{\chi' \cdot \chi''}(G; x)$?

10. The Simple Significance of the Distinguished Character Scheme

Consider the class $\mathfrak{B}$ of connected bipartite plane graphs and the distinguished character scheme $\chi^d(G)$ of these graphs (see (6)). It is convenient to write $D(G; x)$ for $\widetilde{f}_{\chi^d}(G; x)$. Note that $\mathfrak{B} \supset \mathfrak{Z} \supset \mathfrak{H}^0$ where $\mathfrak{Z}$ is the class of connected bipartite plane graphs all of whose faces satisfy $l(F) \equiv 2, \mod 4$, and $\mathfrak{H}^0$ is the class of hexagonal systems (benzenoid graphs) with an even number of vertices (necessarily, v is even for all graphs in $\mathfrak{Z}$): for these classes, $\chi^d(G)$ reduces to $\mathbf{1}$, thus

(7) $$D(G; x) = K(G; x) \quad (G \in \mathfrak{Z}).$$

One might expect that $D(G; x)$ has particularly interesting properties: indeed it has, however it turns out that $D(G; x)$ is essentially the same as $f(G; x)$; precisely:

PROPOSITION 5.

(8)
$$D(G; x) = (-i)^v f(G; ix),$$
$$f(G; x) = (-i)^v D(G; ix) \quad (G \in \mathfrak{B}).$$

To say this in other words, recall that the ordinary spectrum of $G \in \mathfrak{B}$ is real and symmetric, i.e., if y is a (nonzero) eigenvalue of multiplicity m of G then so is $-y$. This implies that $f(G; x)$ has the form

$$f(G; x) = x^r \prod_{j=1}^{s} (x^2 - y_j^2) = x^r \left(x^{2s} - a_2 x^{2s-2} + \cdots + (-1)^s a_{2s} \right) \quad (r \geq 0, r + 2s = v)$$

where the $a_{2\sigma}$ are positive. Then

$$D(G; x) = x^r \prod_{j=1}^{s} (x^2 + y_j^2) = x^r (x^{2s} + a_2 x^{2s-2} + \cdots + a_{2s}),$$

i.e., the distinguished spectrum is obtained from the ordinary spectrum by multiplying all eigenvalues by i, and conversely (see entries 1,2 and 4 in Table 5 where $G \in \mathfrak{Z}$).

PROOF OF PROPOSITION 5. Let G have bipartition $\mathbb{B}$, $\mathbb{W}$. Clearly, the orientation δ "from $\mathbb{B}$ to $\mathbb{W}$" induces the distinguished character scheme of G. Arrange the vertices such that $\mathbb{B}$ comes before $\mathbb{W}$. Then, with some 0, 1-matrix $\mathbf{U}$ of size $|\mathbb{B}| \times |\mathbb{W}|$,

$$\mathbf{A}(G) = \begin{pmatrix} \mathbf{0} & \mathbf{U} \\ \mathbf{U}^\top & \mathbf{0} \end{pmatrix}, \quad \widetilde{\mathbf{A}}(G^\delta) = \begin{pmatrix} \mathbf{0} & \mathbf{U} \\ -\mathbf{U}^\top & \mathbf{0} \end{pmatrix},$$

$$\mathbf{A}^2(G) = \begin{pmatrix} \mathbf{UU}^\top & \mathbf{0} \\ \mathbf{0} & \mathbf{U}^\top\mathbf{U} \end{pmatrix}, \quad \widetilde{\mathbf{A}}^2(G^\delta) = \begin{pmatrix} -\mathbf{UU}^\top & \mathbf{0} \\ \mathbf{0} & -\mathbf{U}^\top\mathbf{U} \end{pmatrix} = -\mathbf{A}^2(G).$$

If $\operatorname{spec} \mathbf{A}(G) = \{y_1, -y_1; \ldots; y_s, -y_s; 0, 0, \ldots, 0\}$ $(y_\sigma > 0)$ then

$$\operatorname{spec} \widetilde{\mathbf{A}}^2(G^\delta) = \operatorname{spec}\left(-\mathbf{A}^2(G)\right) = \{-y_1^2, -y_1^2; \ldots; -y_s^2, -y_s^2; 0, 0, \ldots, 0\};$$

using the fact that the nonzero eigenvalues of $\widetilde{\mathbf{A}}(G^\delta)$ occur in complex-conjugated pairs, we conclude that

$$\operatorname{spec} \widetilde{\mathbf{A}}(G^\delta) = \{iy_1, -iy_1; \ldots; iy_s, -iy_s; 0, 0, \ldots, 0\}$$
$$= i \operatorname{spec} \mathbf{A}(G).$$

Hence the assertion. $\square$

11. A Consequence for the Theory of Benzenoids

By (7) and (8), for any $G \in \mathfrak{Z}$,

$$f(G; x) = (-1)^{v/2} K(G; ix);$$

especially, if G is a benzenoid graph with an even number of vertices, then f and K are related as follows.

$$f(\mathrm{G}; x) = x^r \prod_{j=1}^{s}(x^2 - y_j^2) = x^r\big(x^{2s} - a_2 x^{2s-2} + - \cdots + (-1)^s a_{2s}\big),$$

$$K(\mathrm{G}; x) = x^r \prod_{j=1}^{s}(x^2 + y_j^2) = x^r(x^{2s} + a_2 x^{2s-2} + \cdots + a_{2s}).$$

In molecular-orbital theory (Hückel's model, LCAO-MO theory) of polycyclic hydrocarbons whose graphs (carbon skeletons) are elements of $\mathfrak{Z}$ (e.g., benzenoids with an even number of vertices), $f(\mathrm{G}; x)$ plays the central role: via a couple of ingenious simplifications, it is based on Schrödinger's equation; its zeros (the eigenvalues of $\mathbf{A}(\mathrm{G})$) correspond to the energy levels of the π-electrons of the molecule.

The polynomial $K(\mathrm{G}; x)$ has to do with a much more elementary model, namely, Kekulé's model based on the possible double-bond structures of the molecule (the perfect double-bond structures, also called Kekulé patterns, of the molecule are in $(1, 1)$-correspondence with the perfect matchings of G, their number is $\sqrt{K(\mathrm{G}; 0)}$); the central concept of this model is that of resonance.

The fact that these two polynomials, f and K, turn out to be essentially the same strongly indicates that the two different approaches to the theory of benzenoids—molecular-orbital theory and resonance theory—are indeed closely related.[3]

12. $(3, 6)$-Cages and Their Ordinary Spectra

In concluding let me report an interesting observation which shows that also the ordinary spectrum of a polyhedral graph G (i.e., the family of zeros of $f(\mathrm{G}; x)$) may be of some relevance.

A $(3, 6)$-cage is the graph of a simple polyhedron that has only hexagons and triangles (necessarily, precisely four of them) as its faces; it is regular of degree 3, its number of vertices is a multiple of 4, the smallest is the graph of the tetrahedron.

In 1995, in connection with his investigations of the graphs of fullerene molecules (which are $(5, 6)$-cages), P. W. Fowler (Dept. of Chemistry, Univ. of Exeter) made the intriguing observation that the ordinary spectra $\mathbb{S}$ of the many $(3, 6)$-cages he had investigated all have a very special form, namely,

$$\mathbb{S} = \{3, -1, -1, -1\} \cup \{x_1, x_2, \ldots, x_{2(r-1)}\} \cup \{-x_1, -x_2, \ldots, -x_{2(r-1)}\}$$

where $r = v/4$. Note that $\mathbb{S}_T = \{3, -1, -1, -1\}$ is the spectrum of the tetrahedron; it can easily be proved that $\mathbb{S}_T$ must be contained in $\mathbb{S}$. Thus what remains is the conjecture that the "reduced spectrum" $\mathbb{S} - \mathbb{S}_T$ is symmetric. We have been able to prove this conjecture for all $(3, 6)$-cages on $4q \cdot 2^k$ vertices where k is any nonnegative integer and q is an odd integer between 0 and 100—however, the general case is still open.

This observation may be interpreted as an indication that also for other (sufficiently narrow) classes $\mathfrak{C}$ of polyhedra—in particular, for the class of $(5, 6)$-cages—the ordinary spectrum takes a special form (maybe, not so easily to be detected) reflecting some essential properties of the class $\mathfrak{C}$.

---

[3]The study of the relations between molecular orbital theory and resonance theory has a long history: see, e.g., the work of Dewar/Longuet-Higgins [2] (1952).

References

1. D. M. Cvetković, M. Doob, and H. Sachs, *Spectra of graphs— theory and applications*, 3rd ed., Johann Ambrosius Barth, Heidelberg, 1995.
2. M. S. J. Dewar and H. C. Longuet-Higgins, *The correspondence between the resonance and molecular orbital theories*, Proc. Roy. Soc. London Ser. A **214** (1952), 482–493.
3. P. W. Fowler and D. E. Manolopoulos, *An atlas of fullerenes*, Internat. Ser. Monographs Chem., vol. 30, Clarendon Press, Oxford, 1995.
4. P. W. Fowler, T. Pisanski, and J. Shawe-Taylor, *Molecular graph eigenvectors for molecular coordinates*, Graph Drawing, Lecture Notes in Computer Science, vol. 894, Springer, 1995, pp. 282–285.
5. P. W. Kasteleyn, *Graph theory and crystal physics*, Graph Theory and Theoretical Physics (F. Harary, ed.), Academic Press, London, 1967, pp. 43–110.
6. D. E. Manolopoulos and P. W. Fowler, *Molecular graphs, point groups, and fullerenes*, J. Chem. Phys. **96** (1992), 7603–7614.
7. H. Sachs, P. Hansen, and M. Zheng, *Kekulé count in tubular hydrocarbons*, Match. (1996) no. 33, pp. 169–241.

TECHNISCHE UNIVERSITÄT ILMENAU, INSTITUT FÜR MATHEMATIK, PF 10 05 65, D-98684 ILMENAU, GERMANY

E-mail address: `horst.sachs@mathematik.tu-ilmenau.de`

Centre de Recherches Mathématiques
CRM Proceedings and Lecture Notes
Volume **23**, 1999

The Number of Edge 3-Colourings of the n-Prism

Timothy R. Walsh

During the Open Problems session, Prof. Paul Seymour conjectured that a cubic planar graph with an odd number of edge 3-colourings (up to permutation of the colours) must contain a triangle. That very day, Prof. John Gimbel and I independently proved that the pentagonal prism has 5 edge 3-colourings, thus disproving the conjecture. My proof was of the brute-force variety: I used the obvious greedy algorithm to find all possible edge 3-colourings of the pentagonal prism which fix the colours of the edges incident to a given vertex and constructed the branch-and-bound tree for these colourings (see Fig. 1), and I entered the line-graph of the prism into my program for computing chromatic polynomials [**2**] and verified that the coefficient of $\lambda(\lambda - 1)(\lambda - 2)$ is indeed 5.

Prof. Gimbel's proof was far more elegant—he showed that there is a unique 3-colouring of the edges of a pentagon up to permutation of the colours and rotation, and that each one induces a unique colouring of the entire prism, so that the 5 rotations of the pentagon induce the five colourings of the prism—but he declined to submit his result for publication in the Proceedings. The following theorem generalizes Gimbel's result from the 5-prism to the n-prism.

THEOREM. *The number of edge 3-colourings (up to permutation of the colours) of the n-prism has the same parity as n.*

PROOF. We first show that any 3-colouring of the edges in an n-gon induces a unique edge 3-colouring of the n-prism and any 2-colouring of the edges in an n-gon induces two distinct edge 3-colourings of the n-prism (see Fig. 2). We note first that any colouring of the edges in one of the two n-gons of the prism induces a unique colouring of the side-edges of the prism, given that only three colours are available. If three colours are used on the n-gon E_1, then the side-edges will not all have the same colour, so that there are two consecutive side-edges which are coloured differently. The edge e in the other n-gon E_2 incident to these two side-edges must thus have the same colour as the corresponding edge in E_1. The colouring of e and of the side-edges incident to it determine the colour of the edges

---

1991 *Mathematics Subject Classification.* Primary: 05C15; Secondary: 00A08.

The author wishes to thank Paul Seymour for having made the original conjecture, John Gimbel for having declined to submit his counter-example for publication in the Proceedings, and Pierre Auger, a graduate student at UQAM, for having drawn the figures for this article.

This is the final form of the paper.

TIMOTHY R. WALSH

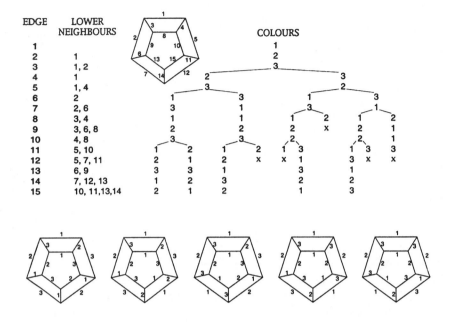

FIGURE 1. The pentagonal prism has exactly five edge 3-colourings fixing the colours of the edges incident to one vertex.

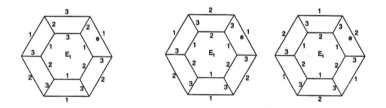

FIGURE 2. A 2-colouring of the edges in the n-gon E_1 induces a unique edge 3-colouring of an n-prism, and a 2-colouring of the edges in E_1 induces two edge 3-colourings of an n-prism.

in E_2 incident to e, so that each of the edges in E_2 must have the same colour as the corresponding edge in E_1. On the other hand, if only two colours are used on the edges in E_1, then all the side-edges will have the same colour, so that one edge e in E_2 can be coloured in two distinct ways, and each of these colourings of e determines the colouring of all the edges in E_2. □

It remains to count in the 3-colourings and 2-colourings of an n-gon. Evidently an n-gon is isomorphic to its own line-graph, so that the edge-colourings of the n-gon can be counted using the chromatic polynomial C_n of the n-cycle. The chromatic polynomial of the n-path is $\lambda(\lambda-1)^{n-1}$. This is the sum of the chromatic polynomial of the n-cycle (formed by joining the end-vertices of the n-path) and that of the $(n-1)$-cycle (formed by identifying the end-vertices of the n-path), so that C_n satisfies the recurrence $C_n + C_{n-1} = \lambda(\lambda-1)^{n-1}$, anchored by the equation

$C_3 = \lambda(\lambda - 1)(\lambda - 2)$. Solving this recurrence yields the formula

$$(1) \qquad C_n = (\lambda - 1)^n + (-1)^n(\lambda - 1)^3 + (-1)^{n-1}\lambda(\lambda - 1)(\lambda - 2).$$

Substituting $\lambda = 3$ into (1), we find that there are $2^n + 2(-1)^n$ colourings of the n-gon with three colours available. Substituting $\lambda = 2$ into (1), we find that there are $1 + (-1)^n$ colourings of the n-gon with two colours available, which is 2 if n is even (1 if we divide by the two permutations of the two colours) and, not surprisingly, 0 if n is odd. If n is odd, then all of the $2^n - 2$ colourings of E_1 with three available colours will actually use all three colours, and each of these colourings induces a unique edge 3-colourings on the n-prism. Dividing by 6, the number of permutations of the three colours, we find that the number of edge 3-colourings of the n-prism, is $(2^{n-1} - 1)/3$, which is odd. If n is even, then of the $(2^{n-1} + 1)/3$ colourings of the n-gon with 3 colours available (up to permutation of the available colours), one of them uses only two colours and the others use all three. The 2-colourings of E_1 induces two distinct edge 3-colourings of the n-prism, and each of the 3-colourings of E_1 induces a unique edge 3-colourings of the n-prims, so that the total number of distinct edge 3-colourings of the n-prism is $(2^{n-1} + 1)/3 + 1$, which is even.

It follows that there are an infinite number of counter-examples to Prof. Seymour's conjecture. It would, of course, be nice to generalize further and conjecture that a non-bipartite cubic planar graph must have an odd number of edge 3-colourings, which would imply the four-colour theorem. But this conjecture has already been shown to be false, since the dodecahedron has ten such colourings [1]. I will resist Prof. Seymour's urgings to make a conjecture lest my conjecture too be proved false with the day.

References

1. H. Sachs, *Invariant polynomials of polyhedra* (these proceedings).
2. T. R. Walsh, *Analysis and optimization of Read's chromatic polynomial algorithm*, Ars Combin. (to appear).

DÉPARTEMENT D'INFORMATIQUE, UNIVERSITÉ DU QUÉBEC À MONTRÉAL, MONTRÉAL (QC), H3C 3P8, CANADA.

E-mail address: `walsh.timothy@uqam.ca`

Centre de Recherches Mathématiques
CRM Proceedings and Lecture Notes
Volume **23**, 1999

The Cost of Radio-Colouring Paths and Cycles

Timothy R. Walsh

Harary [**1**] presented various colour costs of a graph and described the following variation on the chromatic number $\chi(G)$. A "radio-colouring" of a graph G is mapping c from the set of vertices v of G into the set of positive integers such that $d(u, v) = 1$ implies $\left| c(u) - c(v) \right| \geq 2$ and $d(u, v) = 2$ implies $\left| c(u) - c(v) \right| \geq 1$. Let $x(G)$, called the "radio-chromatic" number of G, be the smallest number of colours in which G can be radio-coloured, and let $\Sigma_{i=1...n}c(u_i)$ be the cost of the radio-colouring $c(u_i)$, $i = 1, \ldots, n$. Harary confidently conjectured that the minimum cost of radio-colouring the path P_n is $3n - 2$. We now prove the conjecture in the following equivalent form.

THEOREM 1. *Given a sequence* $a_1, a_2, \ldots, a_n$ *of positive integers satisfying the condition:*

(1) $$|a_i - a_j| \geq 2 \ \text{ if } |i - j| = 1 \quad \text{and} \quad a_i \neq a_j \ \text{ if } |i - j| = 2,$$

the minimum value of $a_1 + \cdots + a_n$ *is* $3n - 2$.

PROOF. We first exhibit, for every positive integer n, a sequence $a_1, a_2, \ldots, a_n$ of positive integers satisfying both (1) and the equation

(2) $$a_1 + \cdots + a_n = 3n - 2.$$

Let $a_1, a_2, \ldots, a_n$ be the sequence

(3) $\quad 1, 4, 2, 5, 1, 4, 2, 5, 1, \ldots, 4, 2, 5, 1$ $\qquad\qquad$ if $n \bmod 4 = 1$,

(4) $\quad 1, 3, 5, 1, 4, 2, 5, 1, 4, 2, \ldots, 5, 1, 4, 2$ $\qquad\quad$ if $n \bmod 4 = 2$,

(5) $\quad 1, 4, 2, 5, 1, 4, 2, 5, 1, \ldots, 4, 2, 5, 1, 4, 2$ $\qquad$ if $n \bmod 4 = 3$,

(6) $\quad 1, 3, 5, 1, 4, 2, 5, 1, 4, 2, \ldots, 5, 1, 4, 2, 5, 1$ $\qquad$ if $n \bmod 4 = 0$.

Obviously, for each n the sequence satisfies (1).

We show by induction on n that it also satisfies (2). For $n = 1$ the sequence (3) is 1 and for $n = 2$ the sequence (4) is $1, 3$; in either case (2) holds. If (2) holds for n it will hold for $n + 2$ because for any value of n the sequence for $n + 2$ is just

---

1991 *Mathematics Subject Classification.* Primary: 05C15; Secondary: 05C38.

The author wishes to thank Prof. F. Harary for posing these problems and making helpful editorial suggestions to improve the presentation of this note.

This is the final form of the paper.

the sequence for n with either $5, 1$ or $4, 2$ appended, and in either case the sum of the two appended terms is 6.

To prove that $3n - 2$ is in fact the minimum, we show that it is a lower bound; that is, every sequence $a_1, a_2, \ldots, a_n$ of positive integers satisfying (1) also satisfies the inequality

$$(7) \qquad\qquad a_1 + \cdots + a_n \geq 3n - 2.$$

To this end we show that any sequence $a_1, a_2, \ldots, a_n$ of positive integers satisfying (1) must end in one of the following six subsequences, each of whose sum is at least 3 times as great as its length, so that if the sequence violates (7) and is longer than its terminal subsequence, then by deleting the subsequence we obtain a shorter sequence which also violates (7).

$$(8) \qquad\qquad a_n, \quad \text{where} \quad a_n \geq 3;$$

$$(9) \qquad\qquad a_{n-1}, 2, \quad \text{where} \quad a_{n-1} \geq 4;$$

$$(10) \qquad\qquad a_{n-1}, 1, \quad \text{where} \quad a_{n-1} \geq 5;$$

$$(11) \qquad\qquad a_{n-2}, 3, 1, \quad \text{where} \quad a_{n-2} \geq 5;$$

$$(12) \qquad\qquad a_{n-2}, 4, 1, \quad \text{where} \quad a_{n-2} \geq 6;$$

$$(13) \qquad\qquad a_{n-3}, 2, 4, 1 \quad \text{where} \quad a_{n-3} \geq 5.$$

If $a_n \geq 3$. then the sequence terminates in subsequence (8). If $a_n = 2$, then by (1) $a_{n-1} \geq 4$, so that the sequence terminates in subsequence (9). If $a_n = 1$, then by (1) $a_{n-1} \geq 3$. Thus, either $a_{n-1} \geq 5$, so that the sequence terminates in subsequence (10), or else $a_{n-1} = 4$ or $a_{n-1} = 3$. If $a_{n-1} = 3$, then by (1) $a_{n-2} \geq 5$, so that the sequence terminates in subsequence (11). If $a_{n-1} = 4$, then by (1) either $a_{n-2} \geq 6$, so that the sequence terminates in subsequence (12), or else $a_{n-2} = 2$. Finally, if $a_{n-2} = 2$, then by (1) $a_{n-3} \geq 5$, so that the sequence terminates in subsquence (13).

It follows that a minimum-length sequence which satisfies (1) but violates (7) must be one of the six subsequences (8)–(13) or else a proper suffix of one of those subsequences. But all of the six subsequences (8)–(13) satisfy (7) and so do all their proper suffixes; thus every sequence which satisfies (1) must also satisfy (7). $\square$

In a private communication, Harary suggested extending these results to the cycle C_n. We show that the minimum value of $\Sigma_{i=1\ldots n}c(u_i)$ for the cycle C_n is $3n$ in the following equivalent form, in which C_{n-1} is represented as P_n with its first and last vertices identified.

THEOREM 2. *Given a sequence* $a_1, a_2, \ldots, a_n$ *of* $n \geq 4$ *positive integers satisfying* (1) *and the additional condition*

$$(14) \qquad\qquad a_n = a_1 \quad \text{and} \quad a_{n-1} \neq a_2,$$

the minimum value of $a_1 + \cdots + a_{n-1}$ *is* $3n - 3$.

PROOF. We first show that any sequence which satisfies (1) and (14) must satisfy the inequality

$$(15) \qquad\qquad a_1 + \cdots + a_{n-1} \geq 3n - 3.$$

We make the minimum colour assigned to a vertex equal to 1 by substracting the same quantity from each of the colours, and then we break C_{n-1} at one of the vertices coloured 1 to turn it into the path P_n. Given any sequence satisfying (1) and (14), let a_m be one of its minimum terms. The sum of the first $n-1$ terms of the sequence $a_m, a_{m+1}, \ldots, a_{n-1}, a_1, a_2, \ldots a_{m-1}, a_m$ is a $a_1 + \cdots + a_{n-1}$. If we substract $a_m - 1$ from each term of the new sequence, all the terms will still be positive and the sequence will still satisfy (1), so that by (7) the sum of all its terms will be at least $3n - 2$. Thus $a_1 + \cdots + a_{n-1} \geq (3n - 2) + n(a_m - 1) - a_m = (3n - 3) + (n - 1)(a_m - 1) \geq 3n - 3$ since $a_m \geq 1$.

To prove that $3n - 3$ is in fact the minimum, we exhibit, for each integer $n \geq 4$, a sequence of positive integers $a_1, a_2, \ldots, a_n$ satisfying (1), (14) and the equation

$$(16) \qquad a_1 + \ldots a_{n-1} = 3n - 3.$$

If $n \bmod 4 = 0$ or 1, we choose the sequence (6) or (3), respectively, both of which satisfy (1) and (14). In the proof of Theorem 1 we have shown that they both satisfy (2), so that since $a_n = 1$ they both satisfy (16). If $n \bmod 4 = 2$, we modify sequence (4) by changing each 1 except a_1 to 2 and each 2 to 1, so that it becomes

$$(17) \qquad 1, 3, 5, 2, 4, 1, 5, 2, 4, 1, \ldots, 5, 2, 4, 1.$$

This sequence satisfies (1) and (14), and by the argument used for (3) and (6) it too satisfies (16). If $n \bmod 4 = 3$ and $n \geq 11$, we insert 3 between the second 1 and the second 5 of (17) so that it becomes

$$(18) \qquad 1, 3, 5, 2, 4, 1, 3, 5, 2, 4, 1, \ldots, 5, 2, 4, 1,$$

which satisfies (1), (14) and (16) as well, since (17) satisfies (16) and we have added 1 to n and 3 to the sum. Finally, if $n = 7$ we choose the sequence $(1, 3, 5, 1, 3, 5, 1)$, which also satisfies (1), (14) and (16). $\qquad \square$

Thus we have shown that the radio-colouring cost of the path P_n is $3n - 2$, and for the cycle C_n it is $3n$, just two more.

References

1. F. Harary, *On color costs of a graph*, private communication.

DÉPARTEMENT D'INFORMATIQUE, UNIVERSITÉ DU QUÉBEC À MONTRÉAL, MONTRÉAL (QC), H3C 3P8, CANADA

E-mail address: walsh.timothy@uqam.ca

Centre de Recherches Mathématiques
CRM Proceedings and Lecture Notes
Volume **23**, 1999

Restricted Graph Coloring: Some Mathematical Programming Models

D. de Werra

ABSTRACT. We consider some coloring problems with additional requirements (restricted coloring problems). We examine in particular Mathematical Programming models for node coloring problems in a graph where each node v must get one color chosen in a set $\phi(v)$ of feasible colors. Additional constraints which are motivated by applications are also considered on cardinalities of color classes. We characterize the cases where the constraint matrix is perfect, balanced or totally unimodular, which would allow a direct approach by linear programming techniques. We review some results in the area and give some extensions and variations.

1. Introduction

There is a collection of scheduling problems which are amenable to graph coloring models. So the attempt to solve scheduling problems via graph theoretical methods has motivated various developments and extensions of graph theory and in particular of graph coloring.

One such extension is the idea of restricted coloring of a graph: it has allowed the introduction of preferences or of preassignment in graph coloring models dedicated to the solution of timetabling or course scheduling problems.

In this paper we shall consider essentially this type of requirement and we will provide mathematical programming formulations of these problems. We will in particular show that in certain cases, integrality properties of the solution can be derived from the structure of the constraint matrix.

In particular we shall characterize in terms of graphs situations where this matrix is balanced or totally unimodular. We shall see that, as could be expected, the graphs for which this matrix is of some of the above types are very special.

The reader is referred to [**2, 3**] for all graph-theoretical terms not defined here.

Some of the results of this paper are mentioned (without proofs) in [**15**]; we give here complete proofs whenever it is appropriate. Additional complexity properties of restricted edge colorings are given in [**15**]. Moreover a collection of examples is

1991 *Mathematics Subject Classification.* Primary: 30A10; Secondary: 30C45.

The author acknowledge support of FCAR (Québec).

This is the final form of the paper.

provided here to illustrate the connections between various results derived in this text.

Our objective here is not to describe in detail polynomial algorithms which can be found elsewhere, but rather to exploit the approach via mathematical programming formulations.

2. Restricted Colorings

Let us first introduce some definitions related to node coloring; for a graph $G = (V, E)$ and a set C of $k = |C|$ colors, a *node k-coloring* of G is an assignment of one color of C to each node in such a way that adjacent nodes have distinct colors.

This concept is often used for solving course scheduling problems: the colors are the periods of the week and courses are represented by the nodes of a graph; conflicting courses (i.e. courses which cannot be scheduled at the same period because they involve the same teacher, or the same students for instance) are linked by edges. A feasible schedule in k periods is then associated with a node k-coloring of the graph.

In most cases, some courses may only be scheduled at one of a prespecified set of periods; this additional requirement may be handled via *restricted node colorings*: for each node $v \in V$, let $\phi(v) \subseteq C$ be a given set of feasible colors for v and let $\phi(V) = \{\phi(v) : v \in V\}$. The restricted node coloring problem (G, ϕ) is to find a node coloring such that each node v gets a color $f(v) \in \phi(v)$.

Since we will give mathematical programming formulations of some coloring problems, we need to recall a few definitions related to matrices.

A $(0, 1)$ matrix is *balanced* if it does not contain a square submatrix of odd order such that each row and each column contains exactly two 1's.

We recall some properties of balanced matrices (see [1, 5]):

If A is a balanced matrix and $\mathbf{b}$, $\mathbf{c}$ nonnegative integral vectors, then the linear programming problem

(2.1)
$$\text{Max } \mathbf{1} \cdot \mathbf{x}$$
$$\text{s.t.} \quad A\mathbf{x} \le \mathbf{b}$$
$$0 \le \mathbf{x} \le \mathbf{c}$$

has an integral solution $\mathbf{x}$.

If A is a balanced matrix, every vertex of

(2.2)
$$\{\mathbf{x} \mid A\mathbf{x} \le \mathbf{1}, \mathbf{x} \ge \mathbf{0}\}$$

has coordinates 0 or 1.

Besides balanced matrices, we shall also refer to *totally unimodular* (t.u.) matrices: a $(0, +1, -1)$ matrix A is t.u. if the determinant of every square submatrix of A has value $0, +1$ or -1.

The following characterization of t.u. matrices was given by Ghouila-Houri [6]:

An integral matrix A is t.u. if and only if any subset J of columns may be partitioned into two subsets J_1, J_2 such that for each row s

(2.3)
$$\left| \sum_{j \in J_1} a_{sj} - \sum_{j \in J_2} a_{sj} \right| \le 1.$$

Furthermore, Hoffman and Kruskal [11] have given this well-known characterization:

An integral matrix A is t.u. if and only if for all integral vectors $\mathbf{d}$, $\mathbf{e}$, $\mathbf{f}$ the polyhedron

(2.4) $$\{\mathbf{x} \mid \mathbf{d} \leq A\mathbf{x} \leq \mathbf{e}, \mathbf{0} \leq \mathbf{x} \leq \mathbf{f}\}$$

has only integral vertices.

For $(0,1)$ balanced matrices, Berge [1] has given a characterization which is analogous to (2.3):

A $(0,1)$ matrix is balanced if and only if any subset J of columns may be partitioned into two subsets J_1, J_2 such that for each row s with

(2.5)
$$\sum_{j \in J} a_{sj} \geq 2, \text{ the following holds}$$

$$\sum_{j \in J_1} a_{sj} \geq 1, \quad \sum_{j \in J_2} a_{sj} \geq 1.$$

This condition shows that a $(0,1)$ t.u. matrix is balanced. Observe that (2.3) and (2.5) may also be formulated for the transposed matrix A^t since by definition A^t is t.u. (resp. balanced) whenever A is t.u. (resp. balanced).

We are now ready to state the restricted coloring problem in terms of the clique-node incidence matrix A of $G = (V, E)$ with $|V| = n$ and $C = \{1, 2, \ldots, k\}$. Here the cliques are supposed to be (inclusionwise) maximal. Following the developments in [8] and [18], we introduce for each color c a vector $\mathbf{x}^c = \big(x(1, c), \ldots, x(n, c)\big)$ such that $x(v, c) = 1$ if node v gets color c and $x(v, c) = 0$ else.

We may now write:

$$\text{Max } z = \mathbf{1} \cdot \mathbf{x}^1 + \cdots + \mathbf{1} \cdot \mathbf{x}^k$$

(2.6)
$$\text{s.t.} \quad A\,x^c \leq \mathbf{1} \qquad (c \in C)$$

$$\sum_{c=1}^{k} x(v, c) \leq 1 \quad (v \in V)$$

$$x(v, c) \in \{0, 1\} \quad (c \in C, v \in V).$$

Observe that (2.6) may be viewed as a problem of finding a maximum stable set of nodes in a graph $\mathcal{G} = G + K_k$ obtained from $G = (V, E)$ by taking k copies $G^1, \ldots, G^k$ of G (G^i has a node set $V_i = \{(v, i) \mid v \in V\}$ and an edge set $E_i = \{[(v, i), (w, i)] : [v, w] \in E]\}$) and constructing a clique on nodes $(v, 1)$, $(v, 2), \ldots, (v, k)$ for each $v \in V$.

We shall denote by $\mathcal{A}(G, k)$ the constraint matrix of (2.6); it is shown in Fig. 1. Now after removing all variables $x(v, c)$ corresponding to pairs (v, c) for which $c \notin \phi(v)$, we may call $B(G, \phi)$ the remaining constraint matrix.

$$\mathcal{A}(G, k) = \begin{bmatrix} A & & & \\ & A & & \\ & & \ddots & \\ & & & A \\ I & I & \cdots & I \end{bmatrix}.$$

FIGURE 1. The constraint matrix of (2.6). A and I appear k times I is the n times n unit matrix where $n = |V|$.

Our problem is now

$$\text{Max } z = \mathbf{1} \cdot \mathbf{x}$$

$$\text{s.t.} \qquad B(G, \phi)\mathbf{x} \le \mathbf{1}$$

(2.7)
$$x \in \{0, 1\}^p$$

$$\text{where } p = \sum_{v \in V} |\phi(v)|.$$

There is a restricted node k-coloring if and only if (2.7) has an optimal $(0, 1)$ solution $\mathbf{x}$ with $\mathbf{1} \cdot \mathbf{x} = |V|$.

Consider now the LP relaxation of (2.7) obtained by replacing the integrality conditions on $\mathbf{x}$ by $\mathbf{x} \ge \mathbf{0}$. Its dual is:

$$\min w = \sum_{K \in \mathcal{K}} \sum_{c \in C} y(K, c) + \sum_{v \in V} y(v)$$

(2.8)
$$\text{s.t.} \qquad \sum_{K : K \ni v} y(K, c) + y(v) \ge 1 \quad \left(c \in \phi(v), v \in V \right)$$

$$y(K, c) \ge 0 \qquad\qquad (c \in C, K \in \mathcal{K})$$

$$y(v) \ge 0 \qquad\qquad (v \in V).$$

Here $\mathcal{K}$ is the family of all (inclusionwise maximal) cliques K of G. Let Γ be a subset of pairs (K, c); we shall say that a node v of G is *covered* by Γ if for each color $c \in \phi(v)$, there is at least one clique K containing node v for which $(K, c) \in \Gamma$. Let $V(\Gamma)$ be the set of nodes of G which are covered by Γ.

A *Hall certificate* of noncolorability for (G, ϕ) is a set Γ of pairs (K, c) such that $|\Gamma| < |V(\Gamma)|$. In [8, 13, 18], characterizations have been given of graphs G with the property that for every possible ϕ, (G, ϕ) has either a restricted coloring or a Hall certificate. Examples of graphs having neither a coloring nor a certificate will be given in the proof of Proposition 3.1.

In the next section, we shall recall some results of [8, 14, 16] and derive refinements and variations of properties given in [15].

Observe that the restricted coloring problem is generally NP-complete [12] since it contains as a special case the node k-coloring problem (when for each node v we have $\phi(v) = \{1, \ldots, k\} = C$).

3. Perfect Matrices and Polyhedral Properties

In fact t.u. and balanced matrices belong to the more general class of *perfect matrices*; these matrices are defined as the clique-node matrices of perfect graphs [3]. It is known that A is perfect if and only if $\{\mathbf{x} \mid A\mathbf{x} \le \mathbf{1}, \mathbf{x} \ge \mathbf{0}\}$ has only integer extreme points. If G is a perfect graph, we may ask when is the restricted k-coloring problem on (G, ϕ) solvable in polynomial time?

It was initially shown in [14] that the following statements are equivalent for a graph G and an integer $k \ge w(G)$ where $w(G)$ is the maximum size of a clique of G

(a) G is a triangulated graph without diamond,
(b) $\mathcal{A}(G, k)$ is perfect for any $k \ge 2$.

This result is proved with different techniques in [**8**, **17**]. In fact, these graphs are precisely the *block graphs*, i.e. the graphs where each block (maximal two-connected component) is a clique.

So let us now recall one of the results in [**8**, **13**] and give some refinements related to perfect graphs.

PROPOSITION 3.1 ([**8**]). *Let G be a simple graph and $C = \{1, 2, \ldots, k\}$ a set of $k \geq 3$ colors. Then the following statements are equivalent*:

1. *G is a block graph,*
2. *$\mathcal{A}(G, k)$ (and hence $B(G, \phi)$) is perfect,*
3. *for every ϕ, (G, ϕ) has either a solution or a Hall certificate.*

As a consequence of this result, the restricted coloring problem can be solved in polynomial time by linear programming: the LP relaxation of (2.7) has a $(0, 1)$ solution $\mathbf{x}$ which is optimal; it can be obtained in polynomial time (see [**9**]). In [**8**] Gröflin gives an algorithm in $O(|V|^4|C|)$ which is combinatorial.

SKETCH OF PROOF.   2. $\Rightarrow$ 3. We shall first as in [**18**] show how a Hall certificate of noncolorability is produced when (G, ϕ) has no solution. The dual (2.8) has a $(0, 1)$ solution $\bar{\mathbf{y}}$ with objective function value $w < |V|$. Let $\Gamma = \{(K, c) \colon \bar{y}(K; c) = 1\}$ and $F = \{v \colon \bar{y}(v) = 0\}$. Then from the dual constraints of the relaxation of (2.7), we have $F \subseteq V(\Gamma)$. Since $w = |\Gamma| + |V - F| < |V|$, we have $|\Gamma| < |F| \leq |V(\Gamma)|$ and Γ is a Hall certificate.

3. $\Rightarrow$ 1. It is furthermore easy to show (see [**18**]) that if G is not a block graph, then there is a ϕ such that (G, ϕ) has neither a solution nor a Hall certificate. Assume G contains a diamond on nodes a, b, c, d; we may construct sets $\phi(v)$ for each node v of G: set $\phi(a) = \{1, 3\}$, $\phi(b) = \{1\}$, $\phi(c) = \{2, 3\}$, $\phi(d) = \{1, 2\}$ as in Fig. 3 and $\phi(v) = \{v\}$ for each $v \neq a, b, c, d$. Clearly there is no coloring of G; there is no certificate Γ: if there exists one, say Γ, we must have $V(\Gamma)$ included in the diamond. We must have two pairs $(K, 2)$ $(L, 3)$ where K and L are 3-cliques because nodes a, c, d have colors 2 and 3 as feasible colors.

Furthermore for color 1, we again must have two pairs $(M, 1)$, $(N, 1)$ where M and N are cliques to cover nodes a, b and d. So we have $|\Gamma| = 4 = |V(\Gamma)|$.

Otherwise if G contains no diamond, let $v_1, v_2, \ldots, v_q$ be the nodes of a chordless cycle Q (of length $q \geq 4$) which must exist if G is not a block graph: assign $\phi(v_1) = \{1\}$, $\phi(v_i) = \{i - 1, i\}$ $(i = 2, \ldots, q - 1)$, $\phi(v_q) = \{q - 1, 1\}$, $\phi(v) = \{v\}$ for each node $v \neq v_1, \ldots, v_q$. It is easy to verify that no restricted coloring of G exists; furthermore if there is a certificate Γ there must be one with $V(\Gamma) \subseteq \{v_1, \ldots, v_q\}$. Now to cover all nodes $v \in Q$ we need $q - 1$ pairs (K, c) for colors $2, \ldots, q$ and at least two pairs (K, c) for $c = 1$ to cover nodes v_1, v_2 and v_q. So there is no Hall certificate.

1. $\Rightarrow$ 2. Finally to show that if G is a block graph, then $\mathcal{A}(G, k)$ is perfect, we refer the reader to [**17**].   □

REMARK 3.1. It is worth observing that for a block graph G with edges ab, ac, ad the matrix $\mathcal{A}(G, 3)$ is perfect according to Proposition 3.1 but not balanced (see Fig. 2).

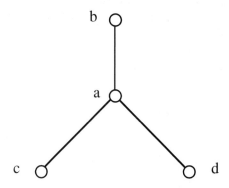

(a) A block graph G with cliques ab, ac, ad.

	a1	b1	c1	a2	c2	d2	a3	b3	d3
ab,1	1	1							
ac,1	1		1						
ac,2				1	1				
ad,2				1		1			
ab,3							1	1	
ad,3							1		1
b		1						1	
c			1		1				
d						1			1

(b) A (9×9) submatrix of $\mathcal{A}(G, 3)$ with row sums and column sums equal to 2.

FIGURE 2. A block graph G for which $\mathcal{A}(G, 3)$ is not balanced.

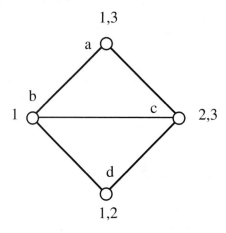

FIGURE 3. A diamond G with feasible sets of colors.

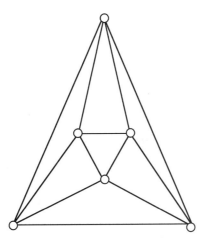

FIGURE 4. A parity graph with a clique node matrix which is not balanced.

The case where $k = 2$ for restricted colorings is a very special case and it may not deserve much attention. We may however examine when $\mathcal{A}(G, 2)$ is perfect or balanced.

PROPOSITION 3.2 ([**14,17**]). *Let G be a simple graph. Then the following statements are equivalent*:

1. *G is a parity graph,*
2. *$\mathcal{A}(G, 2)$ is perfect.*

A graph is a *parity* graph if for any two nodes u, v all chordless chains between u and v have the same parity. These graphs are perfect (see [**4**]). In a parity graph every odd cycle has at least two crossing chords and hence at least two short crossing chords (i.e. chords linking nodes at distance 2 in the cycle) [**4**]. A simple proof using these properties is given in [**16**].

We cannot replace *perfect* by *balanced* in 2. from proposition 3.2: the parity graph in Fig. 4 has a clique-node incidence matrix which is not balanced: it contains a (3×3) submatrix with two ones in each row and in each column.

One may furthermore observe that even for a parity graph G with a t.u. clique node incidence matrix A, the matrix $\mathcal{A}(G, 2)$ may not be balanced: the parity graph G in Fig. 5(a) has four maximal cliques: $\{a, b, c\}$, $\{b, c, d\}$, $\{c, d, e\}$, $\{a, c, e\}$. One checks easily that A is t.u. (since it satisfies (2.3)). However $\mathcal{A}(G, 2)$ contains the (5×5) submatrix given in Fig. 5(b), so it is not balanced.

We may now state the following result which is a refinement of properties given in [**13**] and in [**18**] by restricting the graph G to a line-graph (of a connected graph).

PROPOSITION 3.3. *Let G be a connected line-graph and $C = \{1, 2, \ldots, k\}$ a set of $k \geq 3$ colors. Then the following statements are equivalent*:

1. *G is the line-graph of a tree;*
2. *$\mathcal{A}(G, k)$ (and hence $B(G, \phi)$) is t.u.*
3. *for every ϕ (G, ϕ) has either a solution or a Hall certificate.*

PROOF.   2. $\Rightarrow$ 3. Follows from the proof of Proposition 3.1 and from the observation that $\mathcal{A}(G, k)$ is perfect.

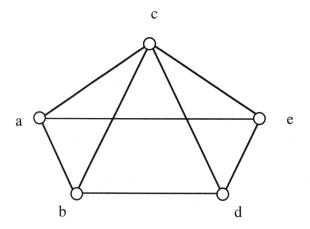

(a) A graph G with a t.u. clique node incidence matrix.

	b1	c1	d1	b2	d2
abc,1	1	1			
cde,1	1	1	1		
bcd,2				1	1
b	1			1	
d			1		1

(b) A (5×5) submatrix of $\mathcal{A}(G, 2)$.

FIGURE 5. A graph G with a t.u. matrix A and $\mathcal{A}(G, 2)$ not balanced.

3. $\Rightarrow$ 1. Suppose G is the line-graph of a (connected) graph H which is not a tree. H contains at least some cycle C of length at least three. If C has length at least four then G also has a chordless cycle of length at least four and the proof of proposition 3.1 applies: we can construct a ϕ such that (G, ϕ) has neither a coloring nor a certificate.

If C has length three, then $H \neq C$ (otherwise, the line-graph of $H = C$ could be considered as well as the line-graph of a $K_{1,3}$ (a claw) and it would be the line-graph of a tree, which contradicts our assumption). So H consists of C with at least one edge $[u, v]$ having exactly one node in common with C (otherwise every edge of H would have both endpoints in C, so G would be a clique and it could be considered as the line-graph of a star $K_{1,n}$). But then G contains a *diamond* (a clique K_4 from which one edge has been removed).

As in the proof of Proposition 3.1, we can construct sets $\phi(v)$ such that (G, ϕ) has neither a coloring nor a certificate.

1. $\Rightarrow$ 2. This was proved in [**13**]; it can alternatively by derived as follows: the line-graph of a tree is a block graph such that each node belongs to at most two

blocks. So each column of the clique node incidence matrix A of G has at most two 1's. So $\mathcal{A}(G,k)$ has at most three 1's in each column.

It was shown in [18] that $\mathcal{A}(G,k)$ is balanced. Now consider $\mathcal{A}(G,k)^t$; it has at most three non zero terms in each row s; $\mathcal{A}(G,k)^t$ satisfies conditions (2.5); they become: any subset J of columns may be partitioned into two subsets J_1, J_2 such that for each row s with $\sum_{j \in J} a_{sj} \geq 2$ the following holds:

$$1 \leq \sum_{j \in J_1} a_{sj} \leq 2, \quad 1 \leq \sum_{j \in J_2} a_{sj} \leq 2.$$

For each row s with $0 \geq \sum_{j \in J} a_{sj} \leq 1$, we have trivially

$$0 \leq \sum_{j \in J_1} a_{sj} \leq 1, \quad 0 \leq \sum_{j \in J_2} a_{sj} \leq 1.$$

So in all cases we have (2.3) and this shows that $\mathcal{A}(G,k)$ is t.u. □

In fact, we can give a stronger statement which characterizes the graphs G for which $\mathcal{A}(G,k)$ is t.u.:

PROPOSITION 3.4. *For a connected graph G, the following statements are equivalent*:

1. *G is the line-graph of a tree,*
2. *$\mathcal{A}(G,k)$ is t.u. for any $k \geq 3$,*
3. *$\mathcal{A}(G,k)$ is balanced for any $k \geq 3$.*

PROOF. We only have to show 3. $\Rightarrow$ 1.

Assume that G is not the line-graph of a tree; if G is a line-graph of a graph H which is not a tree, then G contains either a chordless cycle C with length at least four or a diamond. If G is not a line-graph, then it contains either a $K_{1,3}$ (a claw) or a diamond (see the characterization of line-graphs in [2]).

So we have to consider claws, diamonds and cycles of length at least four.

If G contains a claw, then $\mathcal{A}(G,k)$ contains a 9×9 submatrix with two 1's in each row and in each column (see Remark 3.1).

If G contains a diamond, then $\mathcal{G} = G + K_k$ contains an induced chordless cycle C_7 (see [17]) and $\mathcal{A}(G,k)$ is not perfect (hence not balanced).

If G contains a chordless cycle C of length at least four, then if the length $|C|$ of C is odd, $\mathcal{A}(G,k)$ is not perfect. If $|C| \geq 4$ is even, $\mathcal{G} = G + K_k$ contains an induced C_7 (see [17]). So again $\mathcal{A}(G,k)$ is not perfect. □

4. Application in Chromatic Scheduling

With the above results we are now able to present some applications in chromatic scheduling; consider a graph $G = (V,E)$ whose nodes correspond to items (such as courses in a timetabling problem, or jobs in an open shop scheduling problem). All these items are assumed to have the same duration (say, one period). Each color will be associated with a period.

For each clique K in G we are given a positive integer $b(K)$. A *b-bounded* node k-coloring of G is an assignment of one color in $C = \{1, \ldots, k\}$ to each node such that in each clique K, there are at most $b(K)$ nodes of the same color. If $b(K) = 1$ for each K we have a usual node k-coloring. In timetabling a clique may correspond to courses which cannot all be scheduled simultaneously because they use the same type of classroom (for instance a language lab). If we have $b(K)$ rooms of that type which are available, then a feasible timetable in k periods will correspond to a b-bounded k-coloring.

A restricted b-bounded k-coloring can be defined as before by associating with each node v a set $\phi(v) \subseteq C$ of feasible colors: course v can only be scheduled at periods in $\phi(v)$.

As a consequence of Proposition 3.1, it is possible to determine in polynomial time whether (G, ϕ) has a restricted (usual) k-coloring or not when G is a block graph: The LP formulation has a perfect matrix and according to this, either there exists an optimal integral solution which can be obtained in polynomial time (see [**9**]) or duality will give a type of Hall certificate.

For graphs G such that $\mathcal{A}(G, k)$ is balanced, the same holds for restricted b-bounded k-colorings according to (2.1).

Let us now consider the case where G is a line-graph of some graph H. This means that we are in fact coloring the edges of H. In the case where H is a tree, the constraint matrix of (2.6) is t.u. Let us now define an (a, b)-*bounded* edge k-coloring: it is an assignment of one color c to each edge e in such a way that for each node v, the number $f(v, c)$ of edges of colors c which are adjacent to v satisfies:

$$a(v) \leq f(v, c) \leq b(v),$$

where $a(v)$, $b(v)$ are positive integers satisfying

$$a(v) \leq d(v)/k \leq b(v).$$

Here $d(v)$ is the degree of node v.

Again a restricted (a, b)-bounded edge k-coloring can be defined when each edge e has a set $\phi(e)$ of feasible colors in $\{1, \ldots, k\}$. If $a(v) = 0$, $b(v) = 1$ for each node v, $\phi(e) = C$ for each e, we have a classical edge k-coloring.

If H is a tree, then one can determine in polynomial time whether a restricted (a, b)-bounded k-coloring exists by linear programming: according to Proposition 3.3, the constraint matrix of (2.7) is t.u. and by (2.4), there exists an integral optimal solution or duality will give a noncolorability certificate.

Notice that if $a(v) = \lfloor d(v)/k \rfloor$ and $b(v) = \lceil d(v)/k \rceil$ for each k, we can solve the problem of deciding whether a tree has a restricted *equitable* k-coloring. Such a coloring is used in situations where it is required to balance the loads of teachers and of classes on the k days of the schedule.

For some scheduling applications it may be important to notice when the constraint matrix $\mathcal{A}(G, k)$ is t.u. and not only balanced.

For a restricted (a, b)-bounded node k-coloring, we have a linear programming model when $\mathcal{A}(G, k)$ is t.u.

$$
\begin{aligned}
\max z = \sum_{c \in C} \mathbf{1} \cdot \mathbf{x}^c & \\
\text{s.t. } \mathbf{a} \leq A\mathbf{x}^c \leq \mathbf{b} & \qquad (c \in C) \\
\sum_{c \in C} x(v, c) \leq 1 & \qquad (v \in V) \\
\left.\begin{array}{ll} x(v, c) \geq 0, & c \in \phi(v) \\ x(v, c) = 0, & c \notin \phi(v) \end{array}\right\} & \quad (v \in V)
\end{aligned}
$$

(4.1)

In such a situation we may refine the formulation by considering that for each node v we have not only a set $\phi(v)$ of feasible colors, but also some preferences among these feasible colors: these will be translated by values $g(v,c)$ (given to the assignment of color c to node v). So the objective function in (4.1) may be changed to

$$\text{Max}\, z = \sum_{v \in V} \sum_{c \in \phi(v)} g(v,c) x(v,c).$$

When $\mathcal{A}(G,k)$ is t.u. the resulting problem will still have the integrality properties of (2.4), and it can be solved in polynomial time via linear programming techniques for instance.

At this point, let us mention a relation with the parameter $\alpha_q(G)$ which is the largest possible number of nodes which can be colored using q colors in a graph G. If $\mathcal{A}(G,q)$ is balanced, then the problem can be solved in polynomial time.

Introducing sets of feasible colors $\phi(v) \subseteq C = \{1,\ldots,q\}$ for all nodes v of G, we may ask what is the maximum number $\alpha_q(G,\phi)$ of nodes of G which can be colored with the additional requirements that each node v can only get some color in $\phi(v)$; this is precisely (2.7).

The problem can be solved in polynomial time for instance (with the algorithm of Gröflin [8]) if G is a block graph.

5. Constraints on Cardinalities of Color Sets

In this section we introduce an additional requirement which is sometimes present in chromatic scheduling problems. Given a graph $G = (V, E)$, a set $C = \{1, \ldots, k\}$ of colors and integers $h_1, \ldots, h_k$, we may ask whether there exists a node k-coloring with at most h_c nodes of color c for $c = 1, \ldots, k$. Related results are given in [10].

The problem can now be written as follows:

$$\text{Max}\, z = \mathbf{1} \cdot \mathbf{x}^1 + \cdots + \mathbf{1} \cdot \mathbf{x}^k$$

(5.1)
$$\text{s.t.} \quad \left.\begin{array}{c} A\mathbf{x}^c \le \mathbf{1} \\ \mathbf{1} \cdot \mathbf{x}^c \le h_c \end{array}\right\} \qquad (c \in C)$$

$$\sum_{(c=1)}^{k} x(v,c) \le 1 \qquad (v \in V)$$

$$x(v,c) \in \{0,1\}^{|V| \cdot |C|} \qquad (v \in V, c \in C).$$

It is well known that this problem is NP-complete (even for line-graphs of bipartite graphs (see [16]). Let $\mathcal{D}(G,k)$ be the constraint matrix of (5.1). We may consider the restricted version of this problem by giving to each node v a set $\phi(v)$ of feasible colors. As before let $D(G,\phi)$ be the constraint matrix of the resulting problem. We may again ask for which graphs is the matrix $\mathcal{D}(G,k)$ balanced.

PROPOSITION 5.1. *Let G be a simple graph and $k \ge 2$ an integer. The following statements are equivalent*:

1. *G is a collection of node disjoint cliques,*
2. *$\mathcal{D}(G,k)$ (and hence $D(G,\phi)$) is balanced.*

PROOF.   2. $\Rightarrow$ 1. Assume $G = (V, E)$ is not a collection of node disjoint cliques. Then it contains an induced chain on three nodes a, b, c (i.e. $[a, b]$, $[b, c] \in E$, while $[a, c] \notin E$). Then it is easy to check that $\mathcal{D}(G,2)$ (and hence any $\mathcal{D}(G,k)$)

	⇓ a,1	⇓ b,1	⇓ c,1	⇓ a,2		⇓ c,2
⇒ ab,1	[1]	[1]	0			
⇒ bc,1	0	[1]	[1]			
abc	1	1	1			
				1	1	0
				0	1	1
⇒				[1]	1	[1]
⇒	[1]	0	0	[1]	0	0
	0	1	0	0	1	0
⇒	0	0	[1]	0	0	[1]

FIGURE 6. Matrix $\mathcal{D}(G, 2)$ for a graph G with edges ab, bc.

contains a (5×5) submatrix with exactly two 1's in each row and in each column, so $\mathcal{D}(G, k)$ is not balanced (see Fig. 6).

1. $\Rightarrow$ 2. If G is a disjoint collection of cliques, then its clique node incidence matrix A has a very simple form: its rows are disjoint (no two rows have a 1 in the same column).

Suppose $\mathcal{D}(G, k)$ is not balanced: it must contain a square submatrix P of odd order with exactly two 1's in each row and in each column. P must use at least one row corresponding to a cardinality constraint $\mathbf{1} \cdot \mathbf{x}^c \leq h_c$ (because without these rows, the constraint matrix would be the incidence matrix of the line-graph of a bipartite graph and it could not contain a submatrix P).

Observe that if G consists of exactly one clique, then A reduces to a row of 1's and $\mathcal{D}(G, k)$ is simply the constraint matrix of the transportation problem; so in such a case $\mathcal{D}(G, k)$ does not contain a submatrix P.

So G must consist of at least two cliques. Let $A_1, \ldots, A_k$ be the clique node incidence matrices of G associated to colors $1, 2, \ldots, k$, each one having its additional row of ones. In each A_c for which the additional row is not in P, there is an even number of columns which are used in P.

So there is at least one A_c for which the additional row is used in P and for which an odd number of columns appear in P. Since the additional row of A_c has 1's only in the columns of A_c and since P contains two 1's in each row, there must be at least two columns of A_c in P. Since the number of columns of A_c in P is odd, it is at least three.

But then P contains a row (the additional row of A_c) with at least three non zero entries. This contradicts the assumption on P; hence $\mathcal{D}(G, k)$ is balanced. $\square$

As a consequence, the restricted k-coloring problem with cardinality constraints can be solved in polynomial time when G is a union of node disjoint cliques. Again, combinatorial algorithms could be devised due to the special structure of the graph. In [15] a network flow interpretation is derived for this problem.

Observe also that in Proposition 5.1, $\mathcal{A}(G, k)$ is indeed t.u. (since, as before, it contains at most three 1's in each column).

Final Remarks

Our purpose here was to exhibit and exploit a link between mathematical programming formulations and some solvable cases of generally NP-complete extensions of coloring problems. It is clear that direct approaches (based on combinatorial properties) may be more powerful in the sense that other solvable cases could be obtained.

Furthermore there are many additional variations of restricted coloring problems which should be considered with respect to potential applications. As an example it would be useful to characterize the graphs for which a restricted node k-coloring may be constructed with the additional requirement that there are at least a given number h of nodes with color $1, 2, \ldots, p - 1$ or p (with $p < k$). Such a problem is close to those discussed in [16].

References

1. C. Berge, *Graphes et hypergraphes*, Dunod, Paris, 1970.
2. ———, *Balanced matrices*, Math. Programming **2** (1972), 19–31.
3. ———, *Graphes*, Gauthier-Villars, Paris, 1983.
4. M. Burlet and J.-P. Uhry, *Parity graphs*, Topics on Perfect Graphs (C. Berge and V. Chvátal, eds.), North-Holland Math. Stud., vol. 88, North-Holland, Amsterdam-New York, 1984, pp. 253–277.
5. D. R. Fulkerson, A. J. Hoffman, and R. Oppenheim, *On balanced matrices*, Math. Programming **1** (1974), 120–132.
6. A. Ghouila-Houri, *Caractérisation des matrices totalement unimodulaires*, C. R. Acad. Sci. Paris **254** (1962), 1192–1994.
7. M. Golumbic, *Algorithmic graph theory and perfect graphs*, Academic Press, New York, 1980.
8. H. Gröflin, *Feasible graph coloring and a generalization of perfect graphs*, Inst. for Automation and Operational Research, report no. 199, University of Fribourg, 1992.
9. M. Grötschel, L. Lovasz, and A. Schrijver, *Polynomial algorithms for perfect graphs*, North-Holland Math. Stud., vol. 88, North-Holland, Amsterdam-New York, 1984, pp. 325–356.
10. P. Hansen, A. Hertz, and J. Kuplinsky, *Bounded vertex colorings of graphs*, Discrete Math. **111** (1993), 305–312.
11. A. J. Hoffman and J. B. Kruskal, *Integral boundary points of convex polyhedra*, Linear inequalities and related systems (H. W. Kuhn and A. W. Tucker, eds.), Annals of Math. Stud., vol. 38, Princeton University Press, Princeton, N. J., 1956, pp. 223–246.
12. M. Kubale, *Some results concerning the complexity of restricted colorings of graphs*, Discrete Appl. Math. **36** (1992), no. 1, 35–46.
13. O. Marcotte and P. D. Seymour, *Extending an edge-coloring*, J. Graph Theory **14** (1990), no. 5, 565–573.
14. G. Ravindra and K. R. Parthasarathy, *Perfect product graphs*, Discrete Math. **20** (1977), no. 2, 177–186.
15. D. de Werra, *Restricted coloring models for timetabling*, Discrete Math. **165/166** (1997), 161–170.
16. D. de Werra, J. Blazewicz, and W. Kubiak, *A preemptive open shop scheduling problem with one resource*, Oper. Res. Lett. **10** (1991), no. 1, 9–15.
17. D. de Werra and A. Hertz, *On perfectness of sums of graphs*, Discrete Math. **195** (1999), no. 1-3, 93–101.
18. D. de Werra, A. J. Hoffman, N. V. R. Mahadev, and U. N. Peled, *Restrictions and preassignments in preemptive open shop scheduling*, Discrete Appl. Math. **68** (1996), no. 1-2, 169–188.

Département de Mathématiques, Chaire de Recherche Opérationnelle, École Polytechnique Fédérale de Lausanne (EPFL), CH-1015 Lausanne, Suisse
E-mail address: dewerra@epfl.ch

Centre de Recherches Mathématiques
CRM Proceedings and Lecture Notes
Volume **23**, 1999

Open Problems

PROBLEM 1 (Sylvain Gravier). Let $c(G)$ and $\chi(G)$ be the clique number and the chromatic number of a graph G. If $c(G) < \chi(G)$ does there exist an integer n such that the join of G with a clique on n vertices is such that $c(G+K_n) = \chi(G)+n$?

PROBLEM 2 (Frank Harary). Consider the following two-person game on a 8×8 chessboard where players A and B both have 4 queens: A puts down a queen, B puts down a queen in an independent position (i.e., not taken by the queen of A), A puts down a second queen in an independent position (i.e., not taken by either of the 2 queens already on the board), etc. The last player to put down a queen wins. *Claim*: A can always win.

PROBLEM 3 (Michael Molloy). What is the maximum value of t for which every Δ-regular simple graph has a proper $(\Delta + 1)$-colouring where every vertex has at least t different colours appearing in its neighborhood?

PROBLEM 4 (Horst Sachs). Consider 3 strings of unit length, knotted in some way such that there is no knot between all 3 strings and the set of strings is connected. Given a pair of scissors one may cut a string, but not a knot. *Claim*: one can always create a string of length > 1.

PROBLEM 5 (Horst Sachs). If G is a cubic planar graph with n vertices and a 2-factor which is a union of 2 circuits, then G is edge 3-colourable. This is a special case of the 4-colour theorem; find a shorter proof.

PROBLEM 6 (Paul Seymour). If G is loopless and has no K_{n+1} minor and no induced odd circuit of length ≥ 5 or its complement, then G is n-colourable.